Synthese Library

Studies in Epistemology, Logic, Methodology, and Philosophy of Science

Volume 421

The aim of *Synthese Library* is to provide a forum for the best current work in the methodology and philosophy of science and in epistemology. A wide variety of different approaches have traditionally been represented in the Library, and every effort is made to maintain this variety, not for its own sake, but because we believe that there are many fruitful and illuminating approaches to the philosophy of science and related disciplines.

Special attention is paid to methodological studies which illustrate the interplay of empirical and philosophical viewpoints and to contributions to the formal (logical, set-theoretical, mathematical, information-theoretical, decision-theoretical, etc.) methodology of empirical sciences. Likewise, the applications of logical methods to epistemology as well as philosophically and methodologically relevant studies in logic are strongly encouraged. The emphasis on logic will be tempered by interest in the psychological, historical, and sociological aspects of science.

Besides monographs *Synthese Library* publishes thematically unified anthologies and edited volumes with a well-defined topical focus inside the aim and scope of the book series. The contributions in the volumes are expected to be focused and structurally organized in accordance with the central theme(s), and should be tied together by an extensive editorial introduction or set of introductions if the volume is divided into parts. An extensive bibliography and index are mandatory.

More information about this series at http://www.springer.com/series/6607

Roberto Gronda

Dewey's Philosophy of Science

Springer

Roberto Gronda
Dipartimento di Civiltà e Forme del Sapere
Università di Pisa, Pisa, Italy

Synthese Library
ISBN 978-3-030-37561-4 ISBN 978-3-030-37562-1 (eBook)
https://doi.org/10.1007/978-3-030-37562-1

This Springer imprint is published by the registered company Springer Nature Switzerland AG.
The registered company address is: Gewerbestrasse 11, 6330 Cham, Switzerland

Acknowledgments

This book is the outcome of many years of research. I started working on Dewey's philosophy over a decade ago, when I was an undergraduate at the University of Turin. Not surprisingly, there are a great many people to whom I owe a debt of personal and intellectual gratitude.

First of all, I am grateful to Pierluigi Barrotta for substantially encouraging and supporting my move into the philosophy of science. Without his help, I would never have started this work: he persuaded me that a book on Dewey's philosophy of science was worth writing and provided me with an ideal working environment.

I would like to thank Enrico Pasini, my M.A. advisor. I started working on pragmatism under his supervision, and he taught me pretty much everything I know about reading and pondering a philosophical text.

I have been fortunate to work in the Department of Civilizations and Forms of Knowledge at the University of Pisa. Many of my colleagues spent time discussing various aspects of my work and making invaluable comments. I would like to thank, in particular, Raimondo Cubeddu, Adriano Fabris, Enrico Moriconi, Luca Bellotti, Alfredo Ferrarin, and Alessandra Fussi.

I am also especially grateful to Giacomo Turbanti, Roberto Frega, Torjus Midtgarden, Danilo Manca, Carlo Gabbani, Lorenzo Buccio, Matteo Santarelli, and Lorenzo Azzano. They read and commented on portions of this manuscript and made a number of insightful comments and spotted gaps in the argument.

Many people have greatly assisted me with comments in conversation or writing on different aspects of this material over the past few years. I would especially like to thank my Ph.D. supervisor, Massimo Mugnai, and the members of the committee, Paolo Parrini, Rosa Calcaterra, and James Good. I also thank Giovanni Maddalena, Sarin Marchetti, Anna Boncompagni, Guido Baggio, Gabriele Gava, Tullio Viola, and all the members of the Pragma Group, Jorg Völbers, Mathias Girel, Sami Pihlström, Stéphane Madelrieux, John Shook, David Hildebrand, Jim Garrison, John Capps, Guido Bonino, Paolo Tripodi, Giuliano Guzzone, Alessandro Pagnini, Davide Rizza, Giorgio Lando, Andrea Borghini, and Michele Ginammi.

I would like to thank Luisa for her support, patience and encouragement during the writing of this book.

Most importantly, my gratitude goes to my parents, Orazio and Fulvia, for their love and support through all these years. *Grazie di tutto*.

Introduction

Recent times have witnessed a renewed interest in Dewey's work. In the last few years, much has been written on his philosophies of politics and education, his ethics, and his aesthetics; even his philosophy of religion has been investigated at some depth.[1] Though some sign of interest in it has started to emerge outside pragmatist circles,[2] Dewey's philosophy of science is probably the only relevant

[1]See, for instance, Slater (2008) and Slater (2014, Chapter 4).

[2]The most influential figure in the relatively brief history of the contemporary confrontation of analytic philosophy of science with Dewey's pragmatist approach is Philip Kitcher. See, among others, Kitcher (2011, 2012), in particular Introduction, Chaps. 1, and 5. In Kitcher (2014), Kitcher espouses his views about what counts as pragmatism and acknowledges his theoretical debts toward Dewey's realism and naturalism. In *On Not Being a Pragmatist: Eight Reasons and a Cause*, Ian Hacking denies a direct influence of Dewey's philosophy of science on his own approach; as he puts the matter, "I once tried valiantly to read John Dewey, but it did not click" (Hacking 2007, 45). But see also what Hacking writes in *Representing and Intervening*: "[m]y own view, that realism is more a matter of intervention in the world, than of representing it in words and thought, surely owes much to Dewey" (Hacking 1983, 62). It is hardly questionable that there are some similarities between entity realism and Dewey's "transformational" realism, to the effect that our criteria for ontological commitment have to do with the instrumental capacity to interact with the objects postulated by the best available theory. This point has been lucidly stressed by French and Da Costa in *Science and Partial Truth: A Unitary Approach to Models and Scientific Reasoning* – see French and Da Costa (2003, 177ff.). Contemporary scientific perspectivism seems to have some points in common with Dewey's philosophy of science, in particular in what concerns the conception of naturalism. Ronald Giere labels his own position a "pragmatically inspired naturalism" and insists on his debt toward the pragmatist tradition, in general, and Dewey, in particular Giere (1999, 78) – see also Giere (2006). For an interesting analysis of the possible contributions that a thorough understanding of Dewey's (and Peirce's) philosophy of science can bring to contemporary scientific perspectivism, see Brown (2009, 218ff.). Brown argues that scientific perspectivists – he names Van Fraassen and Giere in his text, but I think that the point he makes holds in general – are concerned with the question of purpose and interest, and yet "the role that purposes play in the processes of representation [...] has been fairly under-specified" (Brown 2009, 218). Now, one of the merits of Dewey's approach is that it provides a sophisticated analysis of the structure of scientific practice, centered around the notion of ends in view. In this sense, Brown holds that contemporary scientific perspectivism can benefit from a confrontation

aspect of his thought to have received no adequate attention from pragmatist scholars. True, there are a few contributions that address specific features of Dewey's theory of inquiry; yet a comprehensive study of Dewey's philosophy of science is lacking.[3] This book aims to fill the gap.

Now, the task of reconstructing Dewey's philosophy of science can be carried out from different perspectives. A purely historical reconstruction can be offered, which aims to highlight and account for the differences between Dewey's views on that topic and those of his contemporaries, as well as for the different phases of his thought.[4] Alternatively, one can opt for a strongly theoretical approach, which purports to assess the theoretical soundness and validity of Dewey's views in light of the most recent research in the field.

It goes without saying that both approaches are valid and essential to a better understanding of Dewey's philosophy of science. Yet in this work, I shall adopt neither of these approaches. The aim of this book is to reconstruct and organize in a systematic and coherent way Dewey's scattered contributions to philosophy of science. In other words, my aim is to present Dewey in his own terms – or, at least, in terms that are as close as possible to his own self-understanding. I have reason to believe this approach is a good choice. On the one hand, I think that a historical work is much needed but that it has to be preceded by a detailed and accurate reconstruction of Dewey's philosophy of science. On the other hand, I believe that any possible translation of the latter into contemporary terms requires the distinctive features of Dewey's position to be ascertained and properly assessed.

In this book, I will therefore present my *semantic* interpretation of Dewey's philosophy of science. "Semantic" here should be understood in a broad way, close to its pre-theoretical usage: in everyday language, it is common to say that a certain thing has a certain meaning for a certain person or that an agent knows what a certain object means, as when it is said that she knows that smoke means fire or that a specific symbol – say, a road sign – means that some action is allowed or prohibited. I use "semantic" as a label to refer to the general fact that something has

---

with pragmatist philosophy of science; I completely agree with him on this point. Other forms of perspectivism are Kantianly inspired: see, in particular, Massimi (2010a, 2018). For a discussion of the strengths and weaknesses of this particular version of perspectivism, see Massimi (2016) and Kitcher's reply, in which he sketches the main outlines of a pragmatist perspectivism which he considers to be theoretically sounder than a Kantian one (Kitcher 2016, 118–119).

[3]The lack of interest in Dewey's philosophy of science is certainly surprising, and yet it is not an isolated incident in pragmatist scholarship. Indeed, with the remarkable exception of Peirce, the concern of classical pragmatists both with science and with the philosophical reflection over science and scientific activity has passed largely unremarked. For a very detailed reconstruction and analysis of the pragmatist tradition in 20th philosophy of science, see Almeder (2007) and Pihlström (2008). In Dewey's case, however, the neglect of Dewey's philosophy of science is particularly bizarre since it is hardly deniable that his theory of inquiry – which is the formula that Dewey uses to refer to what we nowadays call philosophy of science – is central to his entire philosophical project.

[4]Some excellent work has been done in this regard. See, in particular, Richardson (2002), Mormann (2007), Howard (2003), Reisch (2005), and the recent collection of essays Pihlström et al. (2017).

"meaning." From the pragmatist perspective that Dewey endorses, being meaningful means having the capacity to affect and modify the responses that an agent can give to a specific environmental stimulus.

As I read it, Dewey's philosophy of science is built around this fundamental insight. Scientific inquiry is the process through which an agent succeeds in developing highly refined conceptual tools that modify her modes of response and, in so doing, improve and enrich the quality of her transactions with the objects of her environment. This process can be described in two different ways: in naturalistic terms, as the process of transformation of the habits of an organism complex enough to master the use of language, and, in semantic terms, as the process through which the significances of commonsense objects are enriched by the meanings of the concepts that are introduced and articulated within the best scientific theories available. To put it briefly, the idea behind this assumption is that when a scientific theory is accepted – say, contemporary atomic theory – the commonsense object water becomes, in some *specific* contexts, H_2O: the significances of the former are modified and enriched since the agent is now capable of dealing with water in new and more articulated forms.

These two descriptions are substantially equivalent: consequently, by privileging the semantic reading, I by no means intend to minimize or downgrade the importance of Dewey's naturalism. I take the semantic interpretation as somehow more fundamental in this context and for the purposes of this study on the ground that it makes it easier to clarify the role and function played by the different *linguistic* elements that constitute scientific inquiry. Dewey's pragmatist account of scientific concepts, his views of the relationship between science and common sense, his functional account of the elements of inquiry, his notion of scientific objectivity, and his rejection of scientific instrumentalism in favor of a scientific realism revolving around the notion of activity – all these tenets will be shown to depend on Dewey's fundamental semantic insight.

As should be clear from these remarks, Dewey's approach to philosophy of science is strongly idiosyncratic – so much so that the attempt to understand Dewey's philosophy of science on its own terms may seem a task unworthy of pursuit. Accordingly, one may be led to argue that I picked the wrong approach. After all, it might be said, the idea of translation is fascinating: you take something and translate it into a different, more familiar language; as a consequence of that, its philosophical import stands out clearly. So, why not try this approach instead of undertaking an effort at reconstruction of a philosophical proposal in terms that are so strikingly different from the ones that are in use?

The point is that translation is probably the most problematic and controversial activity of all. It presupposes the idea that language is a kind of neutral and inert medium that receives and faithfully transmits any content that may be formulated within it. I reject this assumption – and I particularly reject it in the present case. I believe that there are deep metaphilosophical differences that prevent the translation and direct assimilation of Dewey's views within the framework of much of classical and contemporary analytic philosophy of science.

Here is the basic idea of this book: Dewey's philosophy of science must be framed and understood in the context of his distinctive metaphilosophical outlook; and the reason why many of his theses in the field of philosophy of science may easily look bizarre or unsound – to name a few examples, the idea that propositions are neither true nor false, the idea that inquiry has to do with the reconstruction of objects, or his rejection of the principle of compositionality – is that they are strongly dependent on Dewey's metaphilosophical views. As I use it, metaphilosophy does not simply tell us what philosophy is; more importantly, it tells us what counts as a genuine philosophical problem, as well as what counts as a satisfactory solution to a genuine philosophical problem. It should come as no surprise, therefore, that different metaphilosophical agendas lead to different first-order philosophical theses.

Now, one of the most common moves that I make in this book is that of shifting attention from first-order philosophical theses to Dewey's metaphilosophical commitments and demands. On many occasions, I will ask the reader to change her traditional mindset and try to see things from a different perspective. I am aware that this is something requiring a clarification. The rest of this introduction is dedicated to providing such clarification. In so doing, I will also justify – though mainly indirectly – the approach that I have adopted in this study, as well as its contemporary theoretical relevance.

In their recent and much debated *Pragmatism, Pluralism, and the Nature of Philosophy*, Talisse and Aikin have called "metaphilosophical ascent" the shift of attention from first-order philosophical disputes to metaphilosophical disagreement. They have also stressed that the idea of a metaphilosophical ascent is likely to lead to some sort of antinomy: the more you insist on the dependence of first-order philosophical theses on metaphilosophical stances, the more you run the risk of making philosophical disputes unsubstantial.[5] And they have argued that pragmatists "have tended to place their metaphilosophical agenda at center stage"; that among classical pragmatists, Dewey is the one who is willing to acknowledge the most robust role to metaphilosophy; and that his metaphilosophical impulses are pernicious (Aikin and Talisse 2017, 137).

According to this reading, my insistence on the philosophical relevance of the metaphilosophical assumptions of Dewey's philosophy of science would amount to an admission that his first-order philosophical theses – about the function and nature of science, about scientific objectivity, and about scientific realism – have no value independently of the framework within which they are formulated. In this sense, the metaphilosophical ascent leads to an all-or-nothing approach to first-order philosophical issues: disagreements on first-order philosophical theses are relocated to a higher plane, where no genuine disagreement is possible. Rational

[5]They formulate this point as follows: "The puzzle, then, can be posed as a metaphilosophical antinomy. On the one hand, we seek to accommodate the thought that first-order philosophical programs are manifestations of metaphilosophical stances; on the other, we want to preserve the thought that genuine philosophical disagreement is possible" (Aikin and Talisse 2017, 136).

disagreement – disagreement over specific philosophical theses – is transformed into and replaced with disagreement about what metaphilosophical approach one likes the most. And such disagreement cannot be rationally resolved: it is a matter of personal preference.

I do not agree with Aikin and Talisse on this point, for several reasons. I mention only a few of them. First, I think that their account of the relationship between metaphilosophical assumptions and first-order philosophical theses is too simplistic. They strongly distinguish between metaphilosophical assumptions and first-order philosophical views, without taking into account the fact that the latter can also be understood as the logical articulation of the former. Second, I believe that their conclusion – to the effect that there is no genuine disagreement at the philosophical level – is too rash in that it excludes the possibility that two persons can disagree at the metaphilosophical level while sharing the same first-order philosophical theses. In this sense, Aikin and Talisse's argument is open to the criticism that Laudan leveled against the hierarchical model of rationality (Laudan 1984, 43ff.).

On a more specific level, I do not share their account of Dewey's metaphilosophical commitments. They read Dewey as advancing a "*sociological critique* of traditional philosophy", and they hold that he is not looking for a "new method for doing philosophy nor a new way to resolve traditional philosophical problems", but that he rather aims at a "wholesale reconstruction of philosophy." And they conclude that "philosophy reconstructed along Deweyan lines does not aspire to address traditional problems"; rather the contrary, it gets over these problems and "takes up the wholly new problems deriving from the interface of contemporary democracy and modern science" (Aikin and Talisse 2017, 142).

My interpretation of Dewey's philosophy is more traditional. I concede that the sociological critique of traditional philosophy is a relevant aspect of his thought; I also agree with Aikin and Talisse that some of the problems that Dewey addresses are wholly new, though he couches them in the language of traditional philosophy. For instance, there are good reasons to hold that Dewey's conception of experience is one of these radically new problems, which he invented. Nonetheless, I do believe that these metaphilosophical views are grounded on a more fundamental assumption.

As I read him, Dewey's fundamental metaphilosophical assumption is the idea of the *primacy of activity*. Accordingly, Dewey's genuine metaphilosophical challenge is that of translating traditional problems into a practice-oriented or activity-oriented framework. In so doing, new problems arise; many of these traditional problems are transformed and modified; some others lose their relevance and fade away. But here is the point: one of the theoretical advantages of Dewey's metaphilosophical stance is that it is capable of accommodating and accounting for other metaphilosophical approaches. As a consequence of that, Dewey's philosophy is also capable of accommodating and accounting for first-order philosophical theses that are not in continuity with his metaphilosophy.

This claim is controversial, and I do not want to argue here for its overall validity: there might be parts of Dewey's philosophy that are not amenable to this kind of treatment. But I believe that this is precisely the strategy that Dewey implements in his philosophy of science. An example may be useful here. Consider the *Preface* to *Logic: The Theory of Inquiry*. Here, Dewey writes:

> In the present state of logic, the absence of any attempt at symbolic formulation will doubtless cause serious objection in the minds of many readers. This absence is not due to any aversion to such formulation. On the contrary, I am convinced that acceptance of the general principles set forth will enable a more complete and consistent set of symbolizations than now exists to be made. The absence of symbolization is due, first, to a point mentioned in the text, the need for development of a general theory of language in which form and matter are not separated; and, secondly, to the fact that an adequate set of symbols depends upon prior institution of valid ideas of the conceptions and relations that are symbolized. Without fulfilment of this condition, formal symbolization will (as so often happens at present) merely perpetuate existing mistakes while strengthening them by seeming to give them scientific standing. (LW12, 4)

Here, it is clear that Dewey is not interested in denying the validity of alternative approaches. The point is rather that of putting things in the right perspective and order. A metaphilosophical stance that focuses on language rather than on activity is not fated to develop wrong first-order philosophical theses: as Dewey puts it, there is no aversion on his part to symbolic formulation, and, consequently, there is no aversion – at least in principle – to the consequences that follow from its adoption. Clearly, some of the first-order philosophical theses related to that particular metaphilosophy have to be radically modified, while others have to be abandoned. But this is not what Dewey is concerned with. As I read it, Dewey's criticism is directed against the idea that such an approach can be taken as primitive and fundamental. This is what Dewey cannot accept. The *fundamental* level of analysis – which does not exclude that other, less fundamental levels are possible – is that of activity, and this metaphilosophical option is integral to the first-order thesis, supported by the best scientific evidence available, that human beings are natural organisms transacting with their environment. In this sense, Dewey's metaphilosophy is objectively assessable: it is part of a naturalistic and anti-reductionist image of man in the world. And this is an issue that is of considerable contemporary interest.[6]

Finally, there is another important aspect to which I would like to call attention. Metaphilosophical choices are not a matter of taste; they can be rationally motivated. For instance, it may be shown that some sort of change in metaphilosophical stance is needed because of an explanatory lack in the existing framework of analysis. In my view, the recent practical turn in philosophy of science is a case of this kind.[7] As I read it, the emphasis on the pragmatic dimension of scientific activity

[6]What I have in mind here is Price's distinction between subject naturalism and object naturalism, and the debate that followed. See, in particular, Price (2011), especially Chapter 9, and Price et al. (2015).

[7]The literature on this topic is wide and far from homogeneous. See, in particular, the collections of essays (Nersessian 1987; Pickering 1992; Schatzki et al. 2000; Soler et al. 2014; Agazzi and Heinzmann 2015). See also the influential works of Joseph Rouse, in particular (Rouse 1996, 2002, 2015).

is a consequence of the acknowledgment of the theoretical shortcomings of the theoretical apparatus of traditional philosophy of science.

Now, Dewey's metaphilosophical insistence on the category of activity is certainly at odds with traditional analytic approaches to philosophy of science. As Giere has noticed, an interesting fact in need of explanation is why "a naturalistic pragmatism incorporating an empirical theory of inquiry [got] replaced by a philosophy that regarded induction as a formal relationship between evidence and hypothesis" (Giere 1996, 347). Be that as it may, for quite a long time, philosophers of science working in the latter tradition have largely neglected Dewey's approach, even though many of them – consider, for instance, Ernest Nagel, Patrick Suppes, C. West Churchman, and Richard Rudner – had been influenced by or exposed to it and even though some specific Deweyan theses had been incorporated into the theoretical framework of the discipline.[8]

On the contrary, Dewey's philosophy of science is strongly sympathetic with recent philosophical attempts to understand the nature of scientific practices. This point has been lucidly acknowledged by Hasok Chang who has repeatedly stressed the similarities between his approach and Dewey's views.[9] But Chang is far from being alone in stressing this point: for instance, the same has been done, though from a different perspective, by Arthur Fine (Fine 2007, 59).

I do not want to enter into this area in detail here. My point is a general one. I believe that the affinity between the practical turn in philosophy of science and Dewey's philosophy of science sets the stage for a more accurate and meaningful understanding of the latter. But I would like to take a step further. I may be wrong in this opinion, but I think that the debate on the nature of scientific practices is still largely underdeveloped. The notion of practice is an elusive one: it is not clear

[8]The most relevant of these contributions is Dewey's naturalism. Thanks to Quine, who explicitly approves Dewey's naturalism in *Ontological relativity*, analytic philosophy of science was somehow compelled to take Dewey's pragmatism into serious consideration (Quine 1969, 26–29). Quine's rejection of the analytic/synthetic distinction too can be seen as a pragmatist move of a distinctively Deweyan style. Much has been written on Quine's relationship with pragmatism (see, in particular, Capps (1996) and Godfrey-Smith (2013b)). Barrotta has convincingly shown that the argument advanced by West Churchman and Rudner in support of the thesis that "moral evaluations are inevitably included in the acceptance of a theory" is strongly indebted to pragmatism (Barrotta 2018, 49ff.). It should be clear, therefore, that I am not interested in denying that some sort of continuity exists, on some *specific* themes, between Dewey's pragmatism and the subsequent analytic philosophy of science.

[9]See, among others, Chang (2011b, 211), Chang (2012, 215ff.), Chang (2014, 71), and, in particular, Chang (2016). In the latter text, he writes: "I propose [...] a re-vitalized pragmatist notion of 'coherence', implicit in John Dewey's theory of knowledge, which is irreducible to logical relations between statements. *I define (pragmatist) coherence as a harmonious fitting-together of actions that leads to the successful achievement of one's aims*." The distinctive feature of this notion of coherence is that it "pertains to an activity, not to a set of propositions," in that it is defined "in relation to the aims of the activity in question." Coherence is, therefore, to be understood as "the chief characteristic of a successful activity" (Chang 2016, 112–113). In my view, Chang's shift of attention from propositions to activity is a metaphilosophical move that is very close in spirit to Dewey's metaphilosophical approach.

how to handle it, and it is difficult to find a philosophical apparatus that could satisfactorily account for its distinctive features. In my view, it is difficult even to understand whether or not it has a distinctively *philosophical* use.

In this context, I hold that Dewey's philosophy of science provides a useful theoretical framework for elaborating a consistent account of scientific practices. True, Dewey rarely uses the expression "scientific practice"; he prefers to speak of "secondary experience," "reflection," and "scientific inquiry." And yet Dewey did a lot of work in explaining what a scientific practice is, in clarifying the criteria of identity of scientific practices, in developing metaphilosophical tools for the description and analysis of the normative bonds that tie their elements together, and in elaborating a theory of the ontological commitments of a scientific practice. I am persuaded that all this work, if adequately understood, has a remarkable relevance in the contemporary situation.

It is for this reason that I have chosen to follow a reconstructive approach. Clearly, this choice is far from being theoretically innocent. I do not pretend that my reading is neutral and philosophically unengaged: reconstruction and clarification do not mean or entail theoretical neutrality. So, even though I have tried to stick to Dewey's terminology, and I have abstained from introducing contemporary notions that may lead to a harmful distortion of his views, my interpretation of Dewey's philosophy of science is based on the assumption that the latter is to be read as a philosophy of scientific practices, which is couched in the language of experience and centered around the idea of inquiry as a semantic process of reconstruction and enrichment of the significances of the objects that are used by the agent in her transactions with the world. This should be borne in mind when reading the book.

Plan of the Book

The first two chapters are preliminary. They are dedicated to present those concepts that are necessary in order to understand Dewey's philosophy of science, which is analyzed and assessed in detail in Chaps. 3, 4, and 5.

In Chap. 1, I frame Dewey's philosophy of science in the wider context of his notion of experience. Notoriously, Dewey's account of experience has posed serious difficulties to his interpreters and has been at the center of the quarrel between the defenders of classical pragmatism and the linguistic or analytic neo-pragmatists. My proposal is irenic: I take Dewey's experience to be a function performed by living beings who have the capacity to master the use of language. By conceiving experience as the *locus* of meaning rather than as a kind of object, I focus on the metaphilosophical use that Dewey makes of the notion of experience thus understood, showing that the notion of experience is introduced to account for the plurality of activities or life-behaviors in which human beings are engaged. Finally, I account for the distinctive features of those complex and highly refined activities that Dewey calls "secondary experiences" and that we would call, in our contemporary language, "scientific practices."

In Chap. 2, I address Dewey's philosophy of language. This chapter strengthens and refines the conclusions reached by the first, thus providing a more fine-grained analysis of the structure of life-behaviors. Dewey's theory of inquiry is grounded in the fundamental distinction that Dewey draws between signs and symbols and between significance and meaning. However, that distinction is easily misunderstood. I locate the difference between significance and meaning in the kind of response that signs and symbols solicit, and, in so doing, I explore the relevance of Dewey's semantic externalism and anti-representationalism. Then I proceed to investigate the differences between commonsense and scientific concepts. I pave the way for the formulation of the semantic account of inquiry that I espouse in Chaps. 3 and 4.

Chapter 3 is dedicated to the analysis of the temporal pattern of inquiry. This chapter is mainly metaphilosophical. Indeed, the thesis of the chapter is that the import of Dewey's idea of inquiry should be "cashed out" at a metaphilosophical level, in terms of the language that we should use to describe the structure of scientific activities. More precisely, I argue that Dewey's naturalism, together with his emphasis on the metaphilosophical primacy of the category of activity, puts a number of relevant constraints on what counts as a good philosophical explanation of the different phases that make up scientific inquiries. In this context, I offer a detailed reconstruction of Dewey's concepts of situation and tertiary quality, as well as a new account of his highly controversial notion of warranted assertibility.

In Chap. 4, I discuss the functional elements through which scientific inquiry is carried out. This chapter elaborates on the results reached in Chap. 2. Dewey is explicit that all the elements that are used in inquiry – and which he calls "propositions" – are linguistic, and that taken together they contribute to the construction of the judgment that closes inquiry. More precisely, the function of the existential and the universal propositions – evidence and a priori "knowledge," respectively – is that of constructing the significances and the meanings that will eventually become, when inquiry has reached its satisfactory conclusion, the subject and the object of the final judgment. One of Dewey's most distinctive theses is that the construction of judgment coincides with the construction of scientific objectivity. Indeed, within Dewey's framework, the notion of object is logical: there is no objectivity prior to inquiry. The final part of the chapter is, therefore, devoted to assessing Dewey's pragmatist account of objectivity, with the aim of showing how the latter paves the way for the formulation of a sound form of scientific pluralism.

In Chap. 5, I deal with the issue of scientific realism. Traditionally, Dewey's philosophy of science has been considered as a prototypical form of instrumentalism. I think that this reading is incorrect and that there are good reasons to argue that Dewey is a scientific realist. The problem is rather that of figuring out what kind of scientific realism he embraces and defends in his texts. I reject the interpretation – advanced by Peter Godfrey-Smith – that Dewey's scientific realism is a form of structural realism. Relying on the conclusions reached in the previous chapters, I outline the main features of what I call Dewey's articulative realism, whose key assumption is the idea that the commitment to the existence of the entities postulated by our best scientific theories is a matter of the kind of activities that an agent is capable of undertaking.

A final word is due on what the reader will not find in this book. Dewey's philosophy of science is usually associated with his views about the relations between science and society. So, a book on Dewey's philosophy of science that does not address that issue is likely to be judged incomplete or inadequate. Clearly, I am ready to acknowledge the importance of a careful analysis of this topic for an overall assessment of Dewey's philosophy of science. Nonetheless, I will not deal with it here for a few reasons. First, I believe that Dewey's account of the strong connection existing between scientific activity and the social context in which that activity is carried out is theoretically secondary to his philosophy of science. Consequently, I have preferred to focus on the issue that I consider to be most basic. Second, there already are studies devoted to Dewey's conception of the relations between science and society. For those who are interested in this aspect of Dewey's philosophy, Barrotta's recent book, *Scientists, Democracy and Society: A Community of Inquirers*, says pretty much everything I think should be said about that topic.

Finally, it is indisputable, in my opinion, that Dewey's provocative thesis of an intrinsic connection existing between science and society poses a powerful challenge to contemporary philosophy of science. However, I also think that Dewey's solution is, quite inevitably, out of date: Dewey's world is too different from ours to believe that the conceptual tools he developed to solve *his* problems can be applied to present conditions. Holding the contrary view would be totally unDeweyan. In this case, we do not need a work of conceptual reconstruction and clarification; what we need is, rather, an effort of critical reappraisal and original transformation of Dewey's conceptual apparatus. I have already done some work in this direction.[10] I hope to come up with new and better ideas in the near future.

---

[10]See, in particular, Gronda (2018b) and Barrotta and Gronda (2019).

Contents

List of Abbreviations

Citations of John Dewey's published works are to the 37-volume critical edition published by Southern Illinois University Press under the editorship of Jo Ann Boydston. Citations give series abbreviation, followed by volume number and page number.

Series abbreviations for the Collected Works:

EW The Early Works (1882–1898)
MW The Middle Works (1899–1924)
LW The Later Works (1925–1953)

References to John Dewey's correspondence are to the electronic edition, *The Correspondence of John Dewey*, 1871–1952, Vol. 1–4, edited by L. A. Hickman. Citations give the date of the letter and the names of the sender and the recipient.

References to John Dewey's lectures are to the electronic edition, *Dewey: Lectures* (abbreviated as DL), Vol. 1–2, edited by Donald F. Koch and The Center for Dewey Studies, Southern Illinois University at Carbondale.

Chapter 1
Experience

Abstract This chapter is preliminary. My aim is to frame Dewey's philosophy of science in the wider context of his notion of experience. Dewey's account of experience has posed serious difficulties to his interpreters. My proposal is irenic: I take Dewey's experience to be a function performed by living beings who have the capacity to master the use of language. I focus on the metaphilosophical use that Dewey makes of the notion of experience thus understood, showing that such notion is introduced to account for the plurality of activities or life-behaviors in which human beings are engaged.

Keywords Experience · Semantic identity thesis (SIT) · Life-behavior · Infinity words · Method · Empiricism · Primary/secondary experience · Science and common sense · Experience had/experience known · Cognitive/non-cognitive experience

1.1 Introduction

Experience is probably the most characteristic notion of classical pragmatism. From Peirce to Dewey, through James, experience – and its cognate notion 'empiricism' – are two of the pillars on which the pragmatist approach to knowledge, reality and evaluation relies. Classical pragmatism differs from other relevant contemporary philosophical schools for its attention to the concrete processes through which scientific knowledge is achieved, reality is grasped, and moral and practical evaluations are formulated. A pragmatist-oriented account of knowledge, reality, and evaluation is less concerned with the final product of the process of inquiry than with the process itself: it is the latter which is taken to reveal how an object acquires meaning and significance; which choices are made by the agent, from time to time, in order to accommodate new information within her framework of interpretation; which values are held deer and, consequently, put constraints on the possible courses of action. Experience is the concept that encompasses this complex bundle of practices, and casts particular emphasis on their temporal and contextual dimension.

R. Gronda, *Dewey's Philosophy of Science*, Synthese Library 421,
https://doi.org/10.1007/978-3-030-37562-1_1

In light of the extended and articulated use made by classical pragmatists of the notion of experience, one would therefore expect them to have come up with a clear and consistent view of its nature, structure and function. This should hold particularly true for Dewey, who picks experience as the banner of his philosophical position. Things are rather different, unfortunately. It is quite well known – and many interpreters have stressed this point with strength – that the pragmatist notion of experience is far from being sound. For instance, Richard Rorty has notoriously made a case against the classical pragmatist notion of experience – and, in particular, Dewey's use of it. In his provocative essays *Dewey's Metaphysics* and *Dewey between Hegel and Darwin*, Rorty emphasizes the fact that the notion of experience is more than useless; it is harmful, since it is quite naturally associated with many unfortunate assumptions and insights that pragmatism is willing to counteract: as he explicitly states, "Dewey should have dropped the term 'experience', not redefined" (Rorty 1998, 297).[1] Consequently, he suggests getting rid of it, and rather to focus on language and vocabularies. While speaking of experience brings about only confusion, Rorty remarks, the notions of language and vocabularies are less problematic: while we cannot point to an experience since everything seems to be a good candidate for it, it is quite easy to show what a word is, and it is at least possible in principle to tell a story about the way in which a word relates to other parts of reality – other words or things in the physical world. A similar point has been made by Robert Brandom who, in a footnote of *Articulating Reasons* in which he discusses some terminological and conceptual assumptions of John McDowell's *Mind and World*, states that " 'Experience' is not one of my words": he states that he does not see any "need – either in epistemology or, more important, in semantics – to appeal to any intermediaries between perceptible facts and reports of them that are noninferentially elicited by the exercise of reliable differential responsive dispositions" (Brandom 2000, 206). To many contemporary friends of pragmatism, experience looks like an antique from a bygone past.

One would be tempted to say that Rorty and Brandom's dismissal of the pragmatist notion of experience is largely due to their adoption of a different philosophical agenda, and that such a dismissal alone should not count therefore as evidence against the availability – let alone, the intelligibility – of that notion. This is undoubtedly true, but it is only part of the story. I think it is hard to deny that many of the difficulties to understand the pragmatist account of experience are not primarily concerned with its unsatisfactoriness in light of contemporary philosophical approaches. A large part of these difficulties stems from the intrinsic ambiguity of the notion of experience, which classical pragmatists were unable to

[1]Not surprisingly, Rorty's interpretation of Dewey has been as widely discussed as strongly criticized. For a detailed reconstruction of Rorty's different readings of the Deweyan (kindred) notions of experience and metaphysics see Stuhr (1992). For a criticism of Rorty's account of Dewey's concept of experience, see, among the others, Margolis (2002, 2014). For a 'pragmatist' criticism of Rorty's reading of pragmatism, see Saatkamp (1995). For a comprehensive account of the way in which Dewey's concept of experience has been read and assessed by the so-called neopragmatists, see Hildebrand (2003, 2018).

dissipate. On the one hand, 'experience' is a word full of history, which has been used by many different authors for many different purposes. Classical pragmatists were well aware of its lack of unified meaning, but it is likely that they overestimate their capacity to fix its meaning once for all. This is a failure that Dewey frankly admits in his incomplete 1950 introduction to a new edition of *Experience and Nature* (see LW1, 361). On the other hand, Peirce, James and Dewey do not share a unified, consistent account of experience. In Peirce, experience has mainly to do with the category of Secondness. In some parts of his work, James seems to follow closely Peirce; in others, however, he seems to conceive of experience as a sort of stuff that encompasses subject and object. Dewey, on the contrary, conceives of experience as full-fledged Thirdness, to use Peirce's vocabulary: according to Dewey experience is laden with meanings through and through, and is therefore continuous with what may be called the conceptual dimension or the realm of thought. It seems therefore difficult to trace back all these conceptions of experience to a single, unified account without flattening their differences, and, in so doing, losing the complexity of the notion.[2]

Because of the intrinsic complexity of the subject-matter, the present chapter does not aim to provide a comprehensive account of Dewey's notion of experience. Rather, more modestly, it purports to clarify his conception of experience insofar as it is relevant to his philosophy of science. With this aim in view, I have also decided to keep the references to Dewey's metaphysics to a minimum, and to focus on the role and function that the concept of experience plays in developing and sustaining a pluralistic account of human activities or life-behaviors – or, to use a more contemporary formula, of scientific and common-sense practices.[3] In so

[2]This is one of the reasons for which I have decided, in this chapter, to keep to a minimum any reference to the writings of other pragmatists, in the belief that Peirce and James' accounts of experience do not help shed light on Dewey's one. Clearly, this cautionary assumption does not hold in general; for instance, it is highly reasonable to argue that Peirce's theory of inquiry is explanatory of at least some aspects of Dewey's logic. But I believe that the notion of experience should be treated differently, and more carefully. I do not want to enter into details here, since a satisfactory analysis of this issue would lead us far beyond the limits of the present discussion, but I think that the differences in the understanding of experience are largely due to the different philosophical upbringings of Peirce, James, and Dewey. As is well known, Dewey was much more influenced by Hegel than Peirce and James were, to the extent that, in a well-known autobiographical sketch written in 1930, he remarked that a permanent Hegelian deposit was still acting on his thought (LW5, 154). As said, I will not tackle the controversial issue of Dewey's Hegelianism will since it would lead us astray. Consequently, I will not take a stance on Dewey's faithfulness to Hegel, nor will I attempt to identify the sources of his Hegelianism, which should be traced back to his critical appropriation of the British idealist tradition. More radically, I will not even try to assess whether or not Dewey's philosophy of experience can be labeled as idealistic. This is material for a different kind of work; indeed, all those issues have been extensively debated in recent time: see, among the others, Shook (2000), Good (2006), Shook and Good (2010), Morse (2011), and, for my criticisms of these interpretations, see and Gronda (2011, 2013). However, this particular aspect of Dewey's thought should be always borne in mind while reflecting on his theory of experience.

[3]Dewey's metaphysics is a highly debated and highly problematic issue. For a comprehensive overview of Dewey's metaphysical project, see Boisvert (1988). For a more recent presentation,

doing, by limiting the analysis only to those traits that are not intrinsically connected to Dewey's most controversial views, I hope to achieve two different yet interrelated goals. On the one hand, I would like to make a case for the theoretical viability of Dewey's concept of experience, thus counteracting the widespread view that the latter should be got rid of as an useless conceptual tool. Rather the contrary, I think it can be useful in shedding light on some still undeveloped aspects of the recent practical turn in philosophy of science. On the other hand, I would like to call attention to the metaphilosophical assumptions that lie at the basis of Dewey's attempt to ground his whole philosophy on the notion of experience. Insisting on the metaphilosophical framework in which Dewey articulates his conception of experience should prevent any easy, misleading assimilation of the latter with other, more traditional accounts.

The chapter is divided in four sections. In the first one I lay out what I call the *semantic identity thesis* – namely, the idea that there is no semantic break between objects and our concepts of those objects – and I argue for its compatibility with a strongly naturalistic account of experience. In the second section, I delve deeper into the analysis of Dewey's concept of experience, and highlight its categorial peculiarity, namely, its structural irreducibility to standard object-notions, such as the concept of, say, a chair or a table. In the third section, I take into account Dewey's famous idea that experience should be understood as a method, and suggest that it can be read as a profession of methodological empiricism. Finally, in the fourth section I discuss Dewey's kindred distinctions primary/secondary experience, experience had/experience known, and cognitive/immediate experience, and I argue that they are by no means identical.

1.2 The Semantic Identity Thesis

Dewey struggled to provide a consistent account of experience throughout his entire life. Starting from his early idealistic essays, his philosophical reflections center on understanding the role and function of experience in philosophical inquiry, as well as on elaborating the metaphilosophical resources needed to properly explain them. The two projects go naturally hand in hand: it is because the role and function of experience in philosophical reflection are *sui generis* that a new metaphilosophical apparatus is required.

However, experience is a weasel-word, as Dewey writes in the first edition of *Experience and Nature*, having Ralph Barton Perry in mind. It is weasel since it lends itself to many conflicting interpretations. So, for instance, experience is used

---

see Garrison (2005), Fesmire (2014, chapter 2), Myers (2017) and Alexander (2018). See also Gale (2002) and Myers and Pappas (2004), which is a response to Gale's article. For an analysis of the relation between metaphysics and logic, see Sleeper (2001), in particular Chapter 3 and 5. For a more general discussion about the aims and functions of pragmatist metaphysics, see Pihlstrom (2009) and the two articles Seigfried (2001) and Myers (2004).

to refer to the empirical method through which objective truths about the external world are discovered, as well as to the realm of events that take place within the mind of a subject. Similarly, experience is often understood as something which belongs to an individual – to the extent that, as Dewey remarks, it is almost impossible to speak of experience without being asked "whose experience?", "experience by whom?" (LW1, 178–179) – while, at the very same time, it is commonly held not only that experiences can be repeated, but also that experience is the ultimate ground of intersubjective agreement. The slipperiness and inconsistency of the notion of experience are, therefore, due to the fact that its use usually points in different directions. By the same word, indeed, different authors mean different things; even more troublesome, it may happen that, sometimes, by the same word the very same author means to refer to different and contradictory notions. One of Dewey's main goal is to shed light on such confusion, thus proving that experience is a viable philosophical concept.

Conceptual confusions, however, do not stem out of nowhere; they are always rooted in some features of the subject-matter. In the present case, it is important to note that the slipperiness which the expression 'weasel-word' points at is grounded in the structure of the concept of experience. That notion has an intrinsic dual character which may easily lead astray if one adopts an unsatisfactory metaphilosophical stance. It is such dual character that accounts for most of the confusions that Dewey highlights in his texts. Nonetheless, Dewey maintains, that duplicity should not be taken as a sign of inconsistency. Rather, it is evidence that the notion of experience has to be handled differently from how it has been done in the past by traditional philosophies. The point at stake is precisely to understand how to give a proper account of the notion of experience without trivializing it.

In the second edition of *Experience and Nature*, instead of calling experience a weasel word, Dewey calls it 'double-barreled'. In so doing, he follows the lead of William James who, in his *Essays in Radical Empiricism* remarks that "a given undivided portion of experience" can be taken in two different contexts: in one of these context, it plays "the part of a knower, of a state of mind, of 'consciousness' "; "while in a different context the same undivided bit of experience plays the part of a thing known, of an objective 'content'". For this reason, James explains, "since it can figure in both groups simultaneously we have every right to speak of it as subjective and objective both at once". From those remarks he then concludes that "[t]he dualism connoted by such double-barrelled terms as 'experience,' 'phenomenon,' 'datum,' Vorfindung' – terms which, in philosophy at any rate, tend more and more to replace the single-barrelled terms of 'thought' and 'thing' – that dualism, I say, is still preserved in this account, but reinterpreted, so that, instead of being mysterious and elusive, it becomes verifiable and concrete" (James 1976, 7).

Dewey accepts this Jamesian insight, and formulates it with the following words[4]:

> We begin by noting that 'experience' is what James called a double-barrelled word. Like its congeners, life and history, it includes *what* men do and suffer, *what* they strive for, love, believe and endure, and also *how* men act and are acted upon, the ways in which they do and suffer, desire and enjoy, see, believe, imagine – in short, processes of *experiencing* [...]. It is 'double-barrelled' in that it recognizes in its primary integrity no division between act and material, subject and object, but contains them both in an analyzed totality. 'Thing' and 'thought,' as James says in the same connection, are single-barrelled; they refer to products discriminated by reflection out of primary experience (LW1, 18–19).

The passage is rather complex, and its philosophical implications are wide and far-reaching. I think, however, that the essential core of Dewey's position can be cashed out in purely semantic terms. I take Dewey as saying – amongst the other things, some of which will be discussed and analyzed in the course of this chapter – that in all those cases in which we have no reason to doubt about the validity of our concepts, we also have no reason to doubt that the object constituting the denotatum of the concept is properly described by the properties that the concept predicates – or, stated in other terms, that object and concept share the same content. Object and content, thing and thought, can be distinguished and discriminated only by reflection.[5] So, to give an example drawn from common life, while I am working at my laptop, the object that I am using to type fits perfectly within the modal framework of previsions and expectations established by my concept of laptop. If I pressed the on-off button, the laptop would turn off; if I threw it against the wall, it would likely smash; and so on. The same holds true for scientific objects. If the entities which I am concerned with are atoms of sodium oxygen, I will expect them to behave in a certain way – i.e. to attract electrons more strongly than, let's say,

[4]A clarification here is needed. It has been said that Dewey agrees with James on the double-barrelledness of experience; nonetheless, Dewey's version of James's thesis is more naturalistic oriented and more functionalist. As a consequence of that, I think Dewey's account of the double-barrelled character of experience is heuristically richer than that of James. On the one hand, contrary to James, Dewey provides a simple and satisfactory explanation of the necessary conditions for a given undivided portion of experience to be taken in two different contexts. The functionalist account of the 'division' of experience in different contexts is provided by Dewey's logic of inquiry. On the other hand, Dewey's approach is more naturalistic since Dewey takes the conceptual couples subject/object and consciousness/objective content to be less fundamental than the couple organism/environment. Consequently, from Dewey's perspective, the notion of experience is double-barrelled in the sense that it encompasses both the organism and its environment, while subject and object are two possible configurations that the organism and the environment may take when the organism adopts a logical standpoint. In so doing, the language of epistemology is not taken as a primitive, but contextualized and explained from a genetic and functional point of view.

[5]See, for instance, the following passage from *Experience and Nature*: "Objects are precisely what we are aware of. For objects are events with meanings; tables, the milky way, chairs, stars, cats, dogs, electrons, ghosts, centaurs, historic epochs and all the infinitely multifarious subject-matter of discourse designable by common nouns, verbs and their qualifiers" (LW, 240).

hydrogen; to be partially negatively charged when combined with other atoms, and so on. Their behavior is what their concept predicates of them.

I will call this feature of the double-barrelledness of experience the semantic identity of concept and object.[6] The identity is semantic since what is at stake here is not the existence of the object, but rather its *meaning*, i.e. its significance for the agent that uses it in her life-behavior.[7] Dewey is explicit on this point. It might be argued that, from an epistemic perspective, existence and meaning are so intrinsically interwoven that if we strip any existence of its meanings, nothing is left (see, for instance, EW1, 178–179). However, existence cannot be reduced to meaning since the latter is the outcome of particular transactions between two different kinds of existent entities, the organism and some element of the external world: meaning is a *natural* quality that accrues on some specific, natural transactions between an organism endowed with specific capacities and its environment.

Dewey's use of the notion of meaning is idiosyncratic, and is clearly at odds with the dominant, Fregean tradition in philosophy of language. It is also quite different from the Wittgensteinean well-known thesis that meaning is use, even though some similarities between the two have been detected and emphasized. We will get back to this point later on. For the moment, I would like to draw the attention upon a couple of aspects that are directly related to the semantic identity thesis (henceforth, SIT). First of all, it is worth noting that the idea of a semantic identity of concept and does not entail the impossibility either of error or of discovery. This conclusion would be blatantly false, and would represent a *reductio ad absurdum* of the whole theory. SIT simply articulates the common-sense realist insight that our concepts, *when accepted as valid*, grasp the world in a reliable way. In other terms, SIT states that when our concepts are held to be valid – no matter whether they are scientific or common-sense concepts – we are legitimately committed to the existence of the entities that our concepts postulate, as well as to the properties that our concepts

---

[6]Please note that I do not say that Dewey's idea of the double-barrelledness of experience consists in, or is reducible to, what I have called the semantic identity of concept and object. Rather the contrary, I acknowledge that that idea has many other features that are not semantic – in the sense in which I use this notion. So, for instance, a relevant problem that Dewey addresses in *Experience and Nature* concerns the place of experience in a world of events – which is a distinctively metaphysical issue. My point is more modest: I hold that the semantic identity of concept and object provides a viable starting point for reconstructing Dewey's pragmatist philosophy of science, and that it can be separated – without loss of relevant information – from the other notions with which it is connected in his concept of experience.

[7]A word of clarification is in order. In this chapter I will not distinguish between meaning and significance, and I will use meaning to encompass both the notions, which – starting at least from the *Logic* – Dewey treats as functionally different. This terminological choice is clearly problematic – I concede it – but, in so doing, I conform to the way in which Dewey uses that notion in *Experience and Nature*, as well as in the majority of his works. In any case, a most accurate and fine-grained account of Dewey's notion of meaning is provided in the following chapter, which is devoted to his philosophy of language. I am confident that, taken together, the two chapters succeed in correcting the possible confusions due to the unilaterality of the present exposition.

attribute to them. It does not follow from SIT, thus narrowly conceived, that we cannot be wrong about our existential commitments, nor that we cannot discover something new about the entities we are committed to.[8] SIT is a semantic, not an epistemological, thesis.

Secondly, and consequently, SIT does not support any idealistic conclusion. During his life, Dewey had often been charged with being an idealist, even though of a strange kind. Dewey protested vehemently against such interpretation of his position. He argued that his critics downplayed the importance of the naturalistic element of his account. The naturalistic element is present in the previous quotation, and precisely in the close relation that Dewey draws between experience and life. The same point is formulated with greater clarity in the long essay *The Need for a Recovery of Philosophy* (1917).

In that text, Dewey explicitly states that a satisfactory notion of experience is to be elaborated out of the most recent scientific discoveries in the field of biology. It is clear now, he remarks, that "experiencing means living", and that "[w]here there is life, there is a double connexion maintained with the environment" (MW10, 7). An organism is supported by the environment in its activities, since it has to rely on the natural energies of the environment to successfully perform its acts at the same time, however, living is altering the environment – every action, be it small or great, has some effects on the environment, which change its shape.

The key notion here is that of adaptation. In order to go on living, the organism has to adapt and be adapted to its environment: the adaptation to the objective conditions of the environment is a necessary condition for the conservation of life. This is only part of the story, however. Indeed, Dewey understands the process of adaptation in dynamic terms, thus parting company with the traditional forms of evolutionism, to the effect that the organism is thought to adapt to a fixed and independent environment (LW12, 34–35). On the contrary, Dewey maintains that the idea of "accommodation once for all" is untenable: since the consequences of the activities performed by the organism change the objective environmental conditions, they bring about new forces that disturb and disrupt the previously reached adaptative equilibrium, thus creating new problems that the organism has to face in order to sustain its own life.

Now, according to Dewey, the notion of life centered on the concept of adaptation entails that the two poles of the relation can be separated only *analytically*. It is awkward to say, for instance, that the function of respiration can be accounted for in terms of (a) a set of internal structures (lungs and circulatory system) completed in themselves and independent from the external conditions, which (b) has to be accorded with some elements of the environment. As Dewey writes at

---

[8]As we will see, Dewey has a strong argument to support this view, namely the distinction between primary and reflective experience (see below, Sect. 1.5). However, its plausibility is not dependent on that argument: it is the irreducibility of existence to meaning that warrants the possibility of a sound conception of error and discovery. At the same time, their irreducibility is not understood by Dewey in dualistic terms: if this were so, he would be compelled to embrace the idea of a thing in itself – which is a move that Dewey is not willing to make.

the very beginning of *Human Nature and Conduct* "[b]reathing is an affair of the air as truly as of the lungs; digesting an affair of food as truly as of tissues of stomach", to the extent that it is possible to reverse our traditional way of conceiving our biological activities – according to which, "*we* breathe", "*we* digest" – and rather say that digestion and breathing "are things done by the environment by means of organic structures or acquired dispositions" (MW11, 15). Clearly, there may well be moments in which the adaptive equilibrium between the internal structures and the external conditions is endangered or even lost, but these are moments that become possible only because of the previously existing activity. The function of respiration has therefore a primacy over its constitutive moments. In this sense, Dewey concludes that "[a]ny operative function gets us behind the ordinary distinction of organism and environment", and "presents us with their undifferentiated unity, not with their unification"; indeed, function is "primary" while "distinction is subsequent and derived" (MW13, 377).[9]

The previous remarks are quite uncontroversial, at least from Dewey's perspective. But why should they be relevant to a better understanding of SIT? The point is that Dewey believes that the structural identity that exists between life and experience should make it easy to grasp the untenability of the epistemological dualism that treats subject and object, mind and world, as two separated entities that have to enter into relation to each other in order to bring about experience.

To stress this point, in the Syllabus *Types of Philosophical Thought* (1922–1923) – which contains the preparatory material for *Experience and Nature* – Dewey claims that experience can be identified with a "living function": just as the act of respiration encompasses the physiological and the environmental component, similarly experience encompasses the subjective and the objective factor. The latter case should not be treated as different or as more problematic than the former, in what their structure is concerned. In both cases, the two elements of the function are intimately attuned because they have co-evolved within the context of a specific activity. In this sense, the very notion of adaptation should be better qualified. As Dewey writes, it is important to bear in mind that, properly speaking, adaptation "is *not* adaptation of organism to environment, but adaptation to one another of some of the *many* factors in the function" (MW13, 378). In light of this qualified account of adaptation, it comes as no surprise that our concepts are 'attuned' with their corresponding objects. They are attuned because they have co-evolved as reciprocal

[9]See also what Dewey writes in his early *The Study of Ethics* (1894). "It is the nature of every function to include within itself both organ and environment. The act of respiration is a co-ordination of lungs as organ and air as environment [...]. Function is not the exercise of a predetermined organ upon an external environment, nor is it the adjustment of an organ to a predetermined environment. The nature of the function determines both the organ and the environment. Two animals in whom the function of nutrition is differently performed have, in virtue of that fact, different environments as well as different organs" (EW4, 232–233). The distinction between world and environment is central to Dewey's philosophy. I will deal with it extensively in Chap. 3.

elements within a broader living function, that of experience. And that is precisely what SIT states.

The shift from the elements to the broader context of activity in which those elements are constituted is characteristic of Dewey's thought, and is a distinctive feature of his metaphilosophical approach to philosophical problems. Among the other things, it lies at the ground of Dewey's well known distinction between interaction and transaction, as formulated with regard to the theory of knowledge in *Knowing and the Known*.[10] However, a critic may argue that the whole argument is flawed since one of its premises is ambiguous. Indeed, what does it mean that a living function has a primacy over its constitutive moments? And what kind of primacy do we need in order to justify the conclusion of the argument? Is this primacy explanatory or substantive?[11]

Now, I believe that the explanatory primacy of a function over its constitutive moments can be accepted without great difficulty. It is reasonable to say that, in order to understand what lungs are, one should look at the role they play in the function of respiration. Similarly, even though the parallelism is not complete, in order to understand what the air is, one should look at the properties that make it possible for the air to be part of the process of respiration. Nonetheless, it seems

[10]Here is how Dewey and Bentley understand the notion of transaction, as well as the differences between transaction and interaction: "[o]ur position is that the traditional language currently used about knowings and knowns (and most other language about behaviors, as well) shatters the subjectmatter into fragments in advance of inquiry and thus destroys instead of furthering comprehensive observation for it [. . .]. Our own procedure is the *transactional*, in which is asserted the right to see together, extensionally and durationally, much that is talked about conventionally as if it were composed of irreconcilable separates" (LW16, 67). A few pages later they add that it is possible to distinguish "three levels of the organization and presentation of inquiry", respectively, Self-Action, Interaction, and Transaction. "These levels," they remark, "are all human behaviors in and with respect to the world, and they are all presentations of the world itself as men report it". Those three levels are defined as follows: (1) "*Self-action*: where things are viewed as acting under their own powers"; (2) "*Inter-action*: where thing is balanced against thing in causal interconnection; (3) "*Trans-action*: where systems of description and naming are employed to deal with aspects and phases of action, without final attribution to 'elements' or other presumptively detachable or independent 'entities,' 'essences,' or 'realities,' and without isolation of presumptively detachable 'relations' from such detachable 'elements' (LW16, 100–101).

The last phase of Dewey's philosophy has received relatively little attention from the interpreters, plausibly because of the difficulties in assessing Dewey's specific contribution to the composition of *Knowing and the Known*. One remarkable exception is Frank Ryan who has devoted many of his writings to reconstruct Dewey's late philosophy. See, in particular, Ryan (1997a,b), and, for a comprehensive reconstruction of Dewey's later thought, Ryan (2011).

[11]I use the notion of substantive primacy as a placeholder. Intuitively, the opposite of 'explanatory' should be 'ontological' or 'metaphysical'; the idea is that something can have a primacy for us or in itself. This insight I want to preserve. However, Dewey's argument is clearly not ontological in the usual sense in which that word is now used. Similarly, it is not, strictly speaking, metaphysical, even though it is likely that Dewey would have understood it so. Accordingly, I have decided to use the more neutral formula 'substantive primacy' in order to stress the in-itselfness of the primacy of the function over its component without casting over it an ontological aura that is at odds with Dewey's argument, as well as with his overall approach.

that explanatory primacy is not enough in this context to bear the entire weight of Dewey's argument – for two interrelated reasons.[12] First of all, Dewey does not restrict the analysis to the fact of our epistemic access to the phenomenon. He is clearly dealing with the function in itself, in its objective reality. Secondly, the notion of adaptation is not merely explanatory, unless one is willing to adopt a strong form of anti-realism according to which she is not committed to the existence of the intermediate entities postulated by our best explanation. No matter whether such position be viable or not, this is not what Dewey has in mind. Clearly, Dewey's argument aims to support the conclusion that, in order to know the constitutive moments of a function, the function should be taken as an explanatory primitive. As will see below, this limited conclusion has some relevance for Dewey's understanding of the notion of primary experience, and Dewey sometimes gives the argument a distinctively epistemic turn.[13] That conclusion, however, is not the whole of the argument.[14]

What about the thesis of the substantive primacy? It goes without saying that that is a much stronger assumption. I proceed as follows. The question that I want to address concerns the plausibility of the thesis of the substantive primacy of functions and activities over their constitutive moments. I take the case of breathing, which I consider far less ambiguous than experience. Intuitively, if I can make a case for the substantive primacy of breathing over its external (air) and internal conditions (lungs and circulatory system), I should be able to extend the argument quite easily to the case of experience.

[12]A caveat is needed. By focusing on the difference between the function of experience and that of respiration, one might argue that while explanatory primacy is clearly not enough in the case of respiration, it is sufficient in the case of experience since we are dealing here with a cognitive and epistemic phenomenon. An argument along this line, however, takes for granted something that Dewey openly rejects, namely the possibility of identifying experience with a cognitive phenomenon.

[13]As, for instance, in the following passage from the article *'Consciousness' and Experience*: "digestion, respiration, locomotion express functions, not observable 'objects.' But there is an object that may be described: lungs, stomach, leg-muscles, or whatever. Through the structure we present to ourselves the function; it appears laid out before us, spread forth in detail-objectified in a word. The anatomist who devotes himself to this detail may, if he please (and he probably does please to concentrate his devotion) ignore the function: to discover what is there, to analyze, to measure, to describe, gives him outlet enough. But nevertheless it is the function that fixed the point of departure, that prescribed the problem and that set the limits, physical as well as intellectual, of subsequent investigation. Reference to function makes the details discovered other than a jumble of incoherent trivialities" (MW1, 118–119). See also, on this point, (MW13 381).

[14]The distinction between an explanatory and a properly ontological level has not been adequately emphasized by Gale, who had nonetheless the merit of calling attention to this issue. Discussing Dewey's account of breathing, he rightly remarks that Dewey makes "the relation between organic activities, such as breathing, and its environment mutually internal" (Gale 2010, 64); but see also (Gale 2002, 510). However, he jumps to the conclusion that "Dewey's claim of a mutual dependency between lungs and air, and, more generally, between an organism and its natural environment, is dubious", on the ground that "the air can exist without there being lungs to breathe it and a natural environment can survive the demise of all organisms" (Gale 2010, 65). I think Gale's conclusion is legitimate, but his reconstruction of Dewey's argument is a little bit too easy.

At a first glance, however, such move seems desperate. Even though it may be argued that, without the air making respiration possible, lungs would have not been formed and developed, it seems contrary to facts – and contrary to Dewey's self-professed naturalism – to hold that the function of respiration has a substantive primacy over air. Intuitively, we believe that it is because air was present on Earth that respiration became possible. After all, a world plenty of air but with no breathing animal is far from being unconceivable. So, if this is correct, substantive primacy too will not work as a sound premise for the argument.

However, the idea of substantive primacy can be framed also in different terms.[15] Remember the distinction made above between meaning and existence. Clearly, Dewey's argument does not aim to deny the independent *existence* of the elements of the function; in this sense, the argument is not ontological. Rather the contrary, it aims to deny the possibility that the elements can have a *meaning* independently of their being part of a function or an activity. Take again the function of breathing, and take air. It is indisputable that the air existed and will continue to exist independently of the function of respiration. But it seems equally indisputable that, when entering into the function of respiration as one of its constitutive elements, the air acquires new properties: from those related to the biological processes to which it take parts, up to those connected with the process of phonation.[16] Now, according to Dewey, all these emerging properties cannot be traced back and reduced to the simpler properties of the element since the latter acquires them only as a consequence of its being part of the broader function.[17]

---

[15]I part company here with Eames' reconstruction of Dewey's argument (Eames 1964, 25ff.). I think Eames is right in stressing the importance of the idea of continuity for Dewey's philosophy, as well as in focusing on the intrinsic relation between continuity and emergence. However, I believe that he does not pay enough attention to the semantic nature of Dewey's argument, thus blurring the difference between the semantic and the metaphysical – in Dewey's sense – level of analysis.

[16]See, as an illustration of this point, what Dewey states in the chapter 10 of *Knowing and the Known*. While laying out his notion of transaction, he writes: "[t]his transaction [a trade, which Dewey takes as a good example of what transaction is] determines one participant to be a buyer and the other a seller. No one exists as buyer or seller save *in and because of* a transaction in which each is engaged. Nor is that all; specific things *become* goods or commodities because they are engaged in the transaction. There is no commercial transaction without things which only are goods, utilities, commodities, in and because of a transaction. Moreover, because of the exchange or transfer, both *parties* (the idiomatic name for *participants*) undergo change; and the goods undergo at the very least a change of *locus* by which they gain and lose certain connective relations or 'capacities" previously possessed" (LW16, 242).

[17]This remark is slightly but substantially different from the one made by Kirby in his discussion of Gale's panpsychistic account of Dewey. Kirby states: "just as respiration transforms the object of air chemically, physically, and volumetrically, so an organism transforms its environment. That the natural environment would survive the demise of organism (albeit in a significantly diminished capacity), or the existence of air would continue without lungs, does not negate that the changes made by these respective involvements is not constitutive" (Kirby 2014, 68). Kirby rightly stresses the creativity of the process of adaptation, which entails the production of new objective conditions. On the contrary, my argument is more distinctively semantic, and is concerned with the new

This point should be clearer if we focus on those properties that are not strictly dependent on the physical or chemical makeup of the element. Again, consider air and respiration. Since the process of respiration lies at the basis of the process of phonation, the rejection of the idea of the substantive primacy of the function of respiration over its constitutive moments seems to entail the conclusion that all the properties of air that are displayed in the production of vocal sounds should be taken as somehow existing before, and independently of, the function of respiration. So, for instance, the fact that a negative air pressure must be generated in the lungs in order to pull air into the lungs should be derived from the physical and chemical properties of the element.[18] However, unless one is committed to some sort of strong reductionism about properties, this line of thought should look hardly plausible.[19] It seems more natural to say – at least, this is what Dewey holds, but I think he is right on this point – that these new facts about air and production of sounds stem from the relations into which air enters in the new context provided by the function of respiration.

Let's go back then to experience. If my reading is correct, the thesis of the primacy of experience over its constitutive moments – subject and object, thought and thing, mind and world – should be understood in the following terms. Experience has a primacy over its constitutive moments in that it is constitutively impossible to establish the meaning of an object without taking into account the ways in which the agent deals with it; similarly, it is constitutively impossible to establish the meaning of the actions of the agent without taking into account the objects to which those actions are directed.[20] This insight can be formulated –

properties – and, consequently, the new meanings – that objects acquire as a consequence of their entering into new and original relations with other parts of the environment.

[18] In slightly different terms, Richard Gale has concisely summed up Dewey's argument in favor of what I have called substantive primacy by saying that "it is not only that the lungs require air, but the air that requires lungs" (Gale 2010, 64–65). That statement has a too strong idealistic flavor to be palatable to a realist mind. But it is evident that, when read in terms of meaning rather than existence, it looses great part of its awkwardness. It is far less problematic – and in some ways almost platitudinous, in an emergentist framework as the one which Dewey endorses – to say that air requires lungs in order to acquire the properties related to the functions of respiration and phonation.

[19] It might be replied that, before and independently of the function of respiration, the air has in itself the potentiality to become part of that function. I think that Dewey is sympathetic to the idea of objects as bundle of potentialities that are actualized in and through the different courses of activity in which they can enter. I will discuss Dewey's notion of potentiality in Chap. 4.

[20] I read in this way what Dewey says at the beginning of his famous article *The Postulate of Immediate Experience*. Dewey states that immediate empiricism "postulates that things – anything, everything, in the ordinary or non-technical use of the term 'thing' – are what they are experienced as", so that "if one wishes to describe anything truly, his task is to tell what it is experienced as being". And it is quite clear from the article that what Dewey has in mind is that what a certain object *means* depends on the ways in which the agent (or experiencer) takes it to be. Here is what he says in this regard: "[i]f it is a horse which is to be described, or the *equus* which is to be defined, then must the horse-trader, or the jockey, or the timid family man who wants a 'safe driver,' or the zoologist or the paleontologist tell us what the horse is which is experienced. If these

and, actually, has been formulated – in many different ways: for instance, following Gibson, it can be argued that experience is an ecological event, and that human and animal behavior can only be explained by studying the environment in which that behavior takes place. Or, from a phenomenological point of view, it can be said that experience is characterized by intentionality, where intentionality is the intrinsic aboutness of mind to things. Be that as it may, all those formulations share the same fundamental insight, namely that experience cannot be *accounted for* within an atomistic, bottom-up framework: experience is not the outcome of a process of composition of autonomous and independent elements.

The primacy of experience over its constitutive moments is substantive in that experience is the *locus* – I leave it undetermined for now – in which new properties of the objects *and* new ways of dealing with those objects originate that did not exist before and independently of experience. Experience is, therefore, to be understood as a specific kind of activity – which is made possible by the capacity of the agent to use language – in which meanings are created, which accrue on the existential features of the objects without being reducible to them (Dewey 2012, 321).[21] As the agent undertakes new courses of activity, she comes to experience objects in a new and original way, as a consequence of which objects acquire new and original meanings.

Obviously the acknowledgment of the substantive primacy of experience raises many problems – at different levels of analysis. In particular, it raises some relevant metaphilosophical challenges which have other important philosophical implications: being a function that encompasses subject and object, mind and world, thought and thing as its constitutive moments, the notion of experience is to be handled with great caution. It cannot be accounted for using the vocabularies that are employed to deal with its elements. Dewey is well aware of this fact, to such an extent that, in many of his works, he explicitly stressed the categorial peculiarity of experience. The following section is devoted to the examination of the nature of such categorial peculiarity.

---

accounts turn out different in some respects, as well as congruous in others, this is no reason for assuming the content of one to be exclusively 'real,' and that of others to be 'phenomenal'; for each account of what is experienced will manifest that it is the account of the horse-dealer, or of the zoologist, etc., and hence will give the conditions requisite for understanding the differences as well as the agreements of the various accounts. And the principle varies not a whit if we bring in the psychologist's horse, the logician's horse or the metaphysician's horse" (MW3, 158–159).

[21]The relation between existence and meaning is metaphysical, and is accounted for by Dewey's ontological emergentism. To my knowledge, the best formulation of this position is provided in *Reality as Experience*, where Dewey introduces the notion of *qualitative-transformation-towards*. After reporting the quite obvious objection that "science makes known a chronological period in which the world managed to lead a respectable existence in spite of not including conscious organisms", so that "there was no experience, yet there was reality", Dewey writes that those conditions that antecede experience are "already in transition towards the state of affairs in which they are experience" (MW3, 101). As said I do not want to deal with Dewey's metaphysics in this book; my goal was simply that of stressing the fact that Dewey himself is careful to avoid the strict identification of meaning with existence.

1.3 Experience as Life-Behavior

In the previous section I have analyzed the structure of Dewey's notion of experience, and I have discussed the arguments that he brings in favor of the thesis that experience is a living function. Two issues remain open. The first one concerns the philosophical import of the notion of experience thus conceived. It has been said that experience is double-barrelled. But what does it imply? How should that notion be treated and handled? Secondly, in order to clarify Dewey's argument, I have stressed the similarity between the function of respiration – and, more generally, any biological function – and experience. In doing so, however, it is likely that I ended up flattening the differences between the two kinds of activity. I might have unwittingly conveyed the idea that Dewey takes experience and biological functions as essentially the same, and that experience does not present any distinctive feature that makes it substantially different from, say, the function of respiration or digestion. This is a wrong and misleading impression, which I am going to correct.

Let's start, therefore, with this second issue. In almost all his texts, when it comes down to clarifying the double-barrelledness of experience, Dewey emphasizes its similarity to biological functions such as respiration and digestion. Clearly, this is done for heuristic purposes. As hinted at above, Dewey believes that what he takes to be the indisputable double-barrelled character of biological functions provides strong evidence in favor of the double-barrelledness of experience. At the same time, however, Dewey is ready to take a step further, and sometimes is led to maintain that experience can be *identified*, at least to some extent, with a living function. So, for instance, in the Syllabus *Types of Philosophic Thought* Dewey explicitly states – in a subsection significantly entitled "Experience as Biological" – that "[a]pproaching the matter as externally as possible we may identify experience with a living function" (MW13, 377).

Now, it is worth noting that to say that "[a]pproaching the matter as externally as possible we may identify experience with a living function" does not entail that experience can be reduced to a living function (MW13, 377). The same point is formulated, with greater clarity, in *The Need for Recovery of Philosophy*, where Dewey, after espousing his biological account of the notion of adaptation, states that "[t]hese statements are of an external kind", and they concern "the conditions of experience, rather than [...] experiencing itself" (MW10, 8). What does it mean? Clearly, experience is a function performed by living beings of a particular kind. While a stone does not have an experience, and an ostrich may be said to have experience only metaphorically, human beings are the proper 'agents' or 'bearers' of experience. Their capacity to have an experience is grounded in their biological endowment. However, in order to have an experience, something else is needed. According to Dewey, this additional factor is supplied by their capacity to grasp meanings, which, in the ultimate analysis, is grounded in their capacity to use of language. Rather traditionally, Dewey calls the capacity of grasping meanings 'mind', And yet his conception of mind is idiosyncratic. Here is Dewey's formulation of this point:

> Complex and active animals *have*, therefore, feelings which vary abundantly in quality, corresponding to distinctive directions and phases initiating, mediating, fulfilling or frustrating of activities, bound up in distinctive connections with environmental affairs. They *have* them, but they do not know they have them. Activity is psycho-physical, but not "mental," that is, not aware of meanings. As life is a character of events in a peculiar condition of organization, and "feeling" is a quality of life-forms marked by complexly mobile and discriminating responses, so "mind" is an added property assumed by a feeling creature, when it reaches that organized interaction with other living creatures which is language, communication. Then the qualities of feeling become significant of objective differences in external things and of episodes past and to come (LW1, 198).

The passage is extremely complex since it brings into play so many technical concepts of Dewey's philosophy that a satisfactory discussion of its theoretical import would be impossible in the present context. I will discuss many of those concepts throughout the book, so I simply sketch Dewey's argument. Intuitively, the latter can be expressed as follows. A strong difference exists between animals and human beings. The former have a sentient life which is the outcome of the activities through which they cope with the environment. These activities are constrained partly by their biological endowment and partly by the external conditions that are relevant to their life. The phenomenological episodes that result from the activities in which an animal is engaged – episodes which Dewey calls *feelings* – are qualitatively distinct one from the others, and, as Dewey remarks, correspond to "distinctive directions and phases" of those activities.

However, Dewey argues, animals do have feelings, but they are not aware of them. Strangely, in the above quotation Dewey treats awareness of meanings as a case of knowledge. I take it as a loose way of talking: if simply awareness of meaning were a case of knowledge, that would entail a strong identity between experience and knowledge, a conclusion that Dewey unremittingly and unquestionably rejects. I read Dewey as saying two different, yet strictly interrelated things. First of all, feeling is not enough to have meaning, though the latter is grounded in the former, just as feeling is grounded in the objective phases of activities. Properly speaking, feeling is the distinctive feature of life, while meaning is the distinctive feature of experience, which rests on the capacity of the agent to use language to deal with the features of her environment.[22] Secondly, mind, understood as the capacity to

[22]To my knowledge, the following passage is Dewey's best statement of this fundamental insight. I quote the passage in its entirety: "[l]anguage is a natural function of human association; and its consequences react upon other events, physical and human, giving them meaning or significance. Events that are objects or significant exist in a context where they acquire new ways of operation and new properties. Words are spoken of as coins and money. Now gold, silver, and instrumentalities of credit are first of all, prior to being money, physical things with their own immediate and final qualities. But as money they are substitutes, representations, and surrogates, which embody relationships. As a substitute, money not merely facilitates exchange of such commodities as existed prior to its use, but it revolutionizes as well production and consumption of all commodities, because it brings into being new transactions, forming new histories and affairs. Exchange is not an event that can be isolated. It marks the emergence of production and consumption into a new medium and context wherein they acquire new properties. Language is similarly not a mere agency for economizing energy in the interaction of human beings. It is a

respond to meanings, is a property of natural transactions between the organism and its environment. Mind is therefore defined neither as a substance nor as a function: it is rather a property which is characteristic of the linguistic behavior of human beings in response to the objective conditions of their environment. And clearly, to respond to meanings is not an affair of knowledge, though it has a distinctively cognitive component. As Tiles correctly notices, mind is for Dewey "a label for a system of organic responses of a certain high level of complexity" (Tiles 1990, 80).

Those remarks pave the way for a better understanding of Dewey's well-known formula that experience is a matter of doings and undergoings (MW10, 9; see also MW12, 129). Doings and undergoings must be intrinsically connected – Dewey sometimes states that they must be simultaneous (see MW10, 9), but that term may be misleading – since it is only when an undergoing is perceived to be a consequence of a previous doing that we can speak of experience. Dewey's example of the man asleep who suffers a burn on part of his body – an example formulated in Chapter IV of *Reconstruction of Philosophy* – is intended to highlight precisely this point. In that case, Dewey remarks, "[t]he burn does not perceptibly result from what he has done"; accordingly, the physiological changes in the body do not have any consequence for the subject, or, if they have, "these consequences are not connected with prior doing". In a case like that, therefore, "[t]here is nothing which in any instructive way can be named experience" (MW12, 129). Now, compare it with the case of a child who puts his finger on a fire. It is clear that the act is not the result of a process of reflection; the child acts by instinct. Nonetheless, the undergoing (feeling pain because of the heat) is here connected to the doing (putting the finger on a fire): from that moment one, the latter suggests and means the former, and consequently the child will behave differently when confronted with a fire. The particular quality of the feeling of pain is 'objectified' in becoming a property of a natural event – that the fire burns.[23]

---

release and amplification of energies that enter into it, conferring upon them the added quality of meaning. The quality of meaning thus introduced is extended and transferred, actually and potentially, from sounds, gestures and marks, to all other things in nature. Natural events become messages to be enjoyed and administered, precisely as are song, fiction, oratory, the giving of advice and instruction. Thus events come to possess characters; they are demarcated, and noted. For character is general and distinguished" (LW1, 137–138).

[23]Dewey's example is ambiguous, however, in that it seems to entail the possibility of having an experience independently of the capacity of the child to use the language of the community in which he is raised – a possibility that Dewey explicitly denies in the long passage quoted above, drawn from *Experience and Nature*. I think there is no contradiction here, and that in *Reconstruction in Philosophy*, in which that example is formulated, he simply oversimplifies the argument for explanatory purposes. Indeed, it is clear from the vast majority of his later works, in which an articulated account of meaning is provided, that Dewey is committed to a 'linguistic' conception of meaning. Chapter 2 is entirely devoted to that issue. For the moment, what can be 'cashed out' from the example is the idea that the behavior of an organism dramatically changes when the latter acquires the ability to respond not simply to physical stimuli, but to full-fledged meanings. In doing so, the pattern of behavior of the organism reaches a level of complexity higher than that of a non-linguistic being since it becomes responsive to different and more numerous aspects of the environment.

This example also highlights why the interaction between doings and undergoings is taken by Dewey as constitutive of meaning. The notion of undergoing clearly entails a certain degree of receptivity, but the latter is not to be understood as sheer passivity. It is not enough to undergo a certain set of consequences of physical actions to have an experience of meanings: to count as meanings, these consequences have to be perceived *as* consequences of some specific actions. This is why the man in the example – that is, an organism with linguistic skills – cannot be said to have experience of the burn. Sheer passivity is a degenerate case of receptivity, which lies outside the realm of meaning and experience. Genuine receptivity is possible only within the context of activity, which sets the conditions for the formation of consistent relations of consequences between doings and undergoings. These relations are not merely temporal connections; they are modal relations that support counterfactual inferences of the kind "If I put my finger on a fire, I would burn myself", or "if I had put my finger on that fire, I would have burned myself". That is why these relations are, properly speaking, consistent: they establish a relation between two items that is counterfactually robust.

Now that the difference between life and experience has been clarified, I can move on to tackling the other issue that has remained open, namely, the one concerning the philosophical import of the notion of experience. Dewey deals with what I have called the categorial peculiarity of the notion of experience in several occasions. Its clearest and most articulated formulation is provided in the long introduction to the *Essays in Experimental Logic* (1916).

In that text Dewey focuses on the theoretical status of the notion of experience in order to clarify its role in philosophical discourse. Because of the nature of our language, Dewey remarks, we are used to employ concepts to refer to *objects*. Object-language is more natural to us since we usually employ terms to discriminate relatively permanent elements of our environment – a chair, a table, and so on. In logical terms, the naturality of object-language goes hand in hand with our tendency to frame our understanding of the world in terms of the category of substance, as well as with our disposition to perceive objects as relatively stable configurations of properties. Now, Dewey remarks, the notion of experience is radically different from any 'object-notion' – or, to use his own words, "objective term" (MW10, 324) – and the radical difference between the two lies in that, contrary to the latter, the word 'experience' refers to the broad context in which objects are inserted and with which the experiencer transacts.

Quite interestingly, in the introduction to the *Essays in Experimental Logic* Dewey follows a 'phenomenological' approach which is reminiscent of the one adopted by James in his *Principles of Psychology*. Though it makes use of some notions that are quite controversial or idiosyncratic – such as the distinction between thing and object, or Dewey's functional conception of consciousness – that phenomenological account succeeds well in clarifying the reasons why the notion of experience should be kept distinct from any objective term. I will therefore present Dewey's argument, but I will abstain from discussing it in detail since a careful analysis of its various passages would lead us astray; after that, I will draw some general conclusions that are relevant for my purposes.

Dewey's proto-phenomenological analysis moves from the assumption that the philosophical import of the notion of 'thing' has usually been misunderstood. Contemporary philosophers have indeed tended to equate thing with object, and have considered both thing and object as the correlate of the act of knowing. Within this over-intellectualistic framework of analysis, object is understood as a completely defined element, with clear-cut borders and clear-cut properties.[24] Accordingly, vagueness is boiled down to an epistemic phenomenon, which interferes with the achievement of cognitive goals. The less confused an object of knowledge is, the more appropriate it is to perform its cognitive function.

In reality, Dewey remarks, a thing does not possess the degree of clarity attributed to it by philosophers. In its primary and proper sense, thing is *a res*, *an affair*, something with blurred borders, loaded with expectations and retentions from the past. Every *res*, Dewey writes, "has focus and context[,] brilliancy and obscurity, conspicuousness or apparency, and concealment or reserve, with a constant movement of redistribution" (MW10, 323). Every thing is, therefore, more complex than what we are consciously aware of: in our ordinary transactions with the things that form our environment, we pay attention only to a small part of their features, with the *focus* of our consciousness shifting from time to time, determined by the interest that is predominant in that particular moment.[25]

So, for instance, I am now writing at my desk in my office, typing at the computer: the focus of my attention shifts back and forth from the keyboard to the screen. Ten minutes ago, on the contrary, I was checking a quotation from one of Dewey's books; my computer was obviously there, in the very same position it is now, and yet it was not part of my 'field of consciousness'. It is very likely that I 'noticed' it while reading the book, but I was not truly aware of it; and I was not aware of it because I was not interested in it, just as now I am not aware of the pile of books that I have here on my left, even though I can see them out of the corner of my eye.

Experience is nothing but this broader context which things consist of, and to which objects – as discriminated aspects of the whole 'thing' – belong. The notion is, therefore, introduced to refer to the complex structure of relations, anticipations, expectations, and memories in which an object is embedded, and

---

[24]"Noting things only as if they were objects – that is, objects of knowledge – continuity is rendered a mystery; qualitative, pervasive unity is too often regarded as a subjective state injected into an object which does not possess it, as a mental 'construct,' or else as a trait of being to be attained to only by recourse to some curious organ of knowledge termed intuition" (MW10, 323). I will get back to this point later in Chap. 3, in which I will discuss at length Dewey's notion of situation – which is an infinite or zero word too, as Dewey openly acknowledges (MW10, 324) – and his qualitative account of the normativity of inquiry.

[25]See, in this regard, what Dewey writes in *Knowing and the Known*, that is, in a text written more than 30 years after the *Essays in Experimental Logic*: "*thing* [. . .] is so far from being the metaphysical substance or logical entity of philosophy that is external and presumably physical, that it is 'that with which one is concerned in action, speech, or thought': – three words whose scope not only places *things* in the setting of transactions having human beings as partners, but which [. . .] cover[s] the whole range of human activity" (LW16, 247).

which only gives an object its distinctive meaning. Here is how Dewey articulates this phenomenological insight:

> Movement about an axis persists, but what is in focus constantly changes. "Consciousness", in other words, is only a very small and shifting portion of experience. The scope and content of the focused apparency have immediate dynamic connections with portions of experience not at the time obvious. The word which I have just written is momentarily focal; around it there shade off into vagueness my type writer, the desk, the room, the building, the campus, the town, and so on. In the experience, and in it in such a way as to qualify even what is shiningly apparent, are all the physical features of the environment extending out into space no one can say how far, and all the habits and interests extending backward and forward in time, of the organism which uses the typewriter and which notes the written form of the word only as temporary focus in a vast and changing scene (MW10, 323).

At least four points are worth noting here. (1) The functional nature of consciousness, and its difference from experience. As is clear from the quotation, Dewey conceives of consciousness as the focal moment of experience, and accordingly he denies that the former can be identified with the latter. Dewey's experience is by no means reducible to a philosophy of consciousness, be it of a Cartesian or Husserlian variety. (2) The intrinsic relation between focus and background. Dewey maintains that it is only because there is a background that a focus can emerge. Obviously, background is not held fixed in the sense of being immutable; rather, it is held fixed within a specific context and for a specific purpose. In Dewey's example, the word just written is focal because Dewey is engaged in the activity of type-writing, but it can easily move to the background if the activity changes its direction. This view has many implications for Dewey's theory of inquiry, as we will see in the last section of this chapter. (3) The temporal nature of experience. The equilibrium between focus and background constantly changes. This means that new relations are instituted, and, consequently, new meanings are created in the course of experience. (4) The indefinite extension of experience. While consciousness is focal, experience extends far beyond what is "presently there": it extends in the past and in the future, as well as in the space, and involves external elements (the physical features of the environment) as well as the biological features of the organism (its habits and interests).[26]

It follows, therefore, that the concept of experience is structurally much more complex than objective concepts. Objective terms like chair, desk, laptop, and so on, are used to single out focal parts of an experience. The term 'experience' performs a different function. In ordinary usage, Dewey remarks, experience "denotes something which a specific term like "typewriter" does not designate: namely, the indefinite range of context in which the typewriter is actually set, its spatial and temporal environment, including the habitudes, plans, and activities of its operator" (MW10, 324). In addition, in its constitutive reference to some kind of luminosity, experience suggests "an actual focusing of the world at one point in a

---

[26]The idea of the indefinite scope of experience was criticized and rejected by Russell, who took that idea as a sign of Dewey's idealistically influenced holism, which he deemed as untenable. For a detailed reconstruction of this aspect of Dewey's thought, see Burke (1994).

focus of immediate shining apparency" (MW10, 324). Objective terms denote parts of the experience, but, taken at their face value, they do not bear any reference to the knower or experiencer with whom – please remember SIT – they are essentially related. This is precisely what the notion of experience does, and how it should be used in philosophical discourse. As Dewey openly acknowledges, he creates and introduces the term experience "because of the necessity of having some way to refer peremptorily to what is indicated in only a roundabout and divided way by such terms as 'organism' and 'environment', 'subject' and 'object'; 'persons' and 'things', 'mind' and 'nature', and so on" (MW10, 324).

In order to stress the distinctive denotative function that the notion of experience is expected to carry out in philosophical language, Dewey calls experience an 'infinity' or 'zero' word (MW10, 324). I take Dewey's argument to be close in spirit to Wittgenstein's well known remark, formulated at the beginning of the *Philosophical Investigation*, that language is similar to a tool-box, and that "[t]he functions of words are as diverse as the functions of these objects [hammers, pliers, a saw, a screw-driver, a rule, a glue-pot, glue, nails and screws]" (Wittgenstein 2009, §11). With something of this kind in mind, Dewey stresses the fact that language is constituted of two different kinds of words: terms, which select specific portions of the experiential continuum, and infinity or zero words, which are not terms at all, and whose function consists precisely in reminding that vast and vague continuum that proper terms parse. "I wish simply to ask the reader", Dewey notices, "to bear in mind this radical difference between such words as 'experience', 'reality', 'universe', 'situation', and such terms as 'typewriter', 'me', 'consciousness', 'existence', when used (as they must be used if they are to be terms) in a differential sense" (MW10, 325).

An example is useful to illustrate the different role played by objective terms and zero-words. Take a proper term, like 'rock'. According to SIT, an object is always the object *of* a subject. Consequently, we are entitled to conclude that the object referred to by the term 'rock' means something different to a geologist and a kid that likes to spend his days throwing stones in a lake. Dewey formulates this point, in a slightly different context, by saying that "rocks as minerals signify something more in a group that has learned to work iron than is signified either to sheep and tigers or to a pastoral or agricultural group" (LW12, 66). I read this passage as signifying that the difference in semantic import of the object does not amount to a difference in the reference of the term. Indeed, the objects referred to in the two cases are almost the same.[27] However, the ways in which the two agents – the kid and the geologist – deal with those objects are substantially different: the object enters into different patterns of action and expectation, thus eliciting two radically different behaviors of response. Such bundle of patterns is what meaning consists in from a pragmatist perspective.

[27] I will discuss this point in some detail in the next chapter, Sect. 2.6. For now I simply ask the reader to stick to the common-sense insight that two persons can see, speak and refer to the same object no matter how different their knowledge of it can be.

If this reading is correct, it can be extended to hold in general. Accordingly, it can also be held that the very same object means something different to the same person when it is experienced in different contexts. So, for instance, SIT supports the view that a biologist who has devoted her life to the careful study of snakes would 'see' the snake that she encountered during a walk as a mortal threat rather than as a beautiful specimen of *Vipera Aspis*.[28] The term *Vipera Aspis* can be used, in both situations, to single out the same object – where sameness is, in the case under discussion, sameness of type, not of token – but it is hardly deniable that object gets different meanings, for the same agent, in the two contexts.

It is time to draw these arguments together into a conclusion. First of all, it should now be clear why Dewey believes that a language made up only of proper terms lacks in explanatory power. The point is that the meaning – or, better said, the significance (see Chap. 2) – of the objects which are referred to by proper terms is sensitive to the life-behaviors in which those objects are involved. Consequently, infinity words as experience or situation are needed to remind the fact that subject and object are correlative notions, which presuppose and depend on a broader context for their very possibility. As Dewey remarks, "[i]t is one of the functions of philosophy to recall us from the results of analyses, which are made for special purposes, to the larger, if coarser and in many respects cruder, events which alone have primary existence" (Dewey 2012, 324).

Secondly, the insistence on the substantive primacy of experience is functional to the development of a pluralistic account of life-behaviors. Since an object – as well as the actions of the experiencer dealing with it – acquires its proper significance only within an experience or life-behavior, the possibility of its understanding depends on the acknowledgment of the specific, unique conditions in which *that* particular experience takes place. The world is, therefore, composed of an indefinite plurality of events, some of which are life-behaviors centered on an individual organism that makes use of language to deal with particular traits of an environment at a particular time and for a particular purpose.

Thirdly, the acknowledgment of the indefinite plurality of life-behaviors does not imply that no generalization is possible. Rather the contrary, it is not only possible but also theoretically essential to identify different *types* of experience, on the basis of the distinctive features that they present. In particular, Dewey stresses the difference between primary and reflective experience, remarking that they constitute two radically different ways of dealing with the objects in the world. It is worth noting that primary and reflective experience should not be understood as different

[28]See, in this regard, what Dewey writes in *The Postulate of Immediate Empiricism*: "For example, I start and am flustered by a noise heard. Empirically, that noise *is* fearsome; it *really* is, not merely phenomenally or subjectively so. That *is what* it is experienced as being. But, when I experience the noise as a *known* thing, I find it to be innocent of harm. It is the tapping of a shade against the window, owing to movements of the wind. The experience has changed; that is, the thing experienced has changed – not that an unreality has given place to a reality, nor that some transcendental (unexperienced) Reality has changed, not that truth has changed, but just and only the concrete reality experienced has changed" (MW3, 160).

kinds of sense-data: the primary/reflective experience distinction does not coincide with the primary/secondary quality distinction. They are different modes of practice; different forms of life-behaviors grounded in different vocabularies and different purposes.

I will address this latter issue in Sect. 1.5 and, more extensively, in the next chapter, where I will deal with Dewey's philosophy of language and his theory of concepts. In the next section, I am going to analyze Dewey's metaphilosophical consequences of the idea that experience is an infinity word. In particular, I aim to provide a consistent explanation of Dewey's statement that, in the philosophical language, experience is to be considered as a method, a standpoint from which to properly set up philosophical problems.

1.4 Experience as Method

The first chapter of *Experience and Nature*, significantly entitled *Experience and Philosophic Method*, contains by far the best exposition of Dewey's methodological and metaphilosophical assumptions. Dewey's goal in that chapter is to show that experience and nature are not necessary in conflict, and that their perceived contrast is due to a wrong understanding of those notions rather than to an inner incompatibility between the two. Part of the argument is genuinely metaphysical; part is distinctively metaphilosophical, and relies on conceptual clarification. I will focus exclusively on the latter side of Dewey's argument.

In that chapter Dewey is concerned with the theoretical status of the *philosophical* notion of experience. The starting point of his analysis is the recognition of the problematic character of that concept. Not every notion of experience is problematic. For instance, Dewey argues, *scientific* experience – that is, experience as it is appealed to in science – is taken as an indubitable presupposition of scientific research. Nobody would be willing to deny that natural sciences are and have to be empirical, in the broad sense of being dependent on experience for the ultimate confirmation of their hypotheses. Similarly, nobody would be willing to argue that experimental method shuts us off from nature. Even an anti-realist about scientific objects would not be as radical as to deny that experiments are useful to discover empirical regularities between observables; her point is simply to reject any ontological commitment to the unobservable entities postulated by our best theories.

Dewey is explicit that science is the most reliable way to achieve knowledge about reality, and that the cognitive success of science depends – even though not exclusively – on the confirmatory function performed by experiments. Natural sciences provide the clearest example of a non-conflicting relation between experience and nature. Consequently, they also provide empirical evidence in favor of SIT, since the lack of conflict between experience and nature is a restatement, in different terms, of the idea that there is nothing in principle preventing the possibility of a complete accord between concepts and their respective objects.

Now, Dewey's fundamental assumption is that the *philosophical* concept of experience should be modeled on the idea of *scientific* experience. This thesis has a couple of remarkable metaphilosophical consequences. On the one hand, it entails a strong methodological continuity between science and philosophy. Dewey holds that philosophical reflection is not constitutively different from scientific reflection; for this reason, what holds for the latter should also hold for the former. Since scientific experience is not understood as a veil that prevents us from investigating nature, philosophical experience should be conceived of in the same way. The idea of experience as a set of mental states that have somehow to be accorded with the external world is a theoretical assumption that was originally supported by the best psychological knowledge available, but has now been superseded by the conclusions reached by a biologically oriented psychology. When such an approach to mental phenomena is adopted, the viability of traditional dualistic psychology is radically questioned, and, consequently, the theoretical soundness of any philosophical view grounded in that psychology is fatally undermined. This leads directly to a strong form of philosophical naturalism.

On the other hand, the analysis of how the notion of experience is employed in concrete scientific practice shows that experience is understood as a method instead than as an object.[29] This is clearly compliant with Dewey's distinction between proper terms and infinity or zero words mentioned above. If experience is an infinity word, it follows that its referent cannot be an object, and that a different approach is needed in order to account for the use of that concept. But, apart from Dewey's terminological considerations, are there any reasons in support of such a restrictive assumption? And, more precisely, does scientific experience support it?

Dewey has a strong argument in favor of that assumption. Let's concede for a moment that experience, instead of being a method, is the subject-matter of science. Since experience is by stipulation coincident with reality – the idea of science is grounded in the rational hope, to use Peirce's words, that no aspect of reality lies essentially beyond our grasp – it includes everything. Then, on the assumption that experience is the subject-matter of science – of one *single* science – the latter should bear the burden to provide an account of every aspect of reality, which seems to be a too demanding and hardly-defensible position. Dewey's criticism can be read as a particular version of the slingshot argument: if experience is the subject-matter of a science, then if the latter purports to say something true about some facts of the world, it has to formulate a theory of every possible aspect of reality.[30]

---

[29]Much has been written on this issue. One of the most influential interpretations is Alexander (2004). Of the same author see Alexander (1987). See also Browning (1998) and Garrison (2005).

[30]For the sake of honesty, it has to be said that Dewey's argument is slightly different from the one that I have just presented. His argument, as formulated in the first chapter of the 1925 edition of *Experience and Nature*, centers on philosophical rather than on scientific experience. Here is what Dewey writes in this regard: " "[t]he objection is that experience is then made so inclusive and varied as to be useless for philosophic purposes. Experience, as we are here told to conceive it, includes just everything and anything, actual or potential, that we think of and talk about [...]. [T]he whole wide universe of fact and dream, of event, act, desire, fancy and meanings, valid or

An objector would probably reply to this argument by noticing that what is commonly meant with the idea of experience as the subject-matter of science is that particular features or sections of experience are made object of investigation, not experience taken at large. But this is precisely Dewey's point. What Dewey wants to stress is precisely that any distinction drawn within experience cannot be based on a specific quality of experience *qua* experience: "the notion of experience", he remarks, is "devoid of differential subject-matter – since it includes all subject-matters" (LW1, 373–374). It has therefore to be based on its distinctive content. It is obviously possible to distinguish between the behavior of gases at extremely high pressure, which is part of the subject-matter of kinetic molecular theory, and the behavior of a cell, which is studied by cytobiology. However, in order to single out these subject-matters we need proper terms referring to some stable configurations of the environment. Reference to experience alone is not enough fine-grained to draw those distinctions.

In order to clarify Dewey's position, it may be useful to compare his use of the philosophical notion of experience to Bas van Fraassen's notion of empirical stance. In his *The Empirical Stance*, van Fraassen portraits his empiricism as a philosophical stance. His whole argument relies on the acknowledgment that stances should not be confused with sets of theses. It is quite natural to identify a certain philosophical position with some set of distinctive statements. Van Fraassen calls this idea the (meta-philosophical) Principle Zero.

> **Principle Zero**:
> For each philosophical position X there exists a statement X+ such that to have (or take) position X is to believe (or decide to believe) that X+ (van Fraassen 2002, 41)

Clearly, in the case of empiricism, E+ (the distinctive empiricist statement) has to provide the basis for the criticisms that empiricists raise against metaphysics. In addition, it must be a factual thesis, since no tautology would fit for that role. However, van Fraassen's argument goes, if E+ is to be factual, than the contraries of E+ are legitimate since all those statements are empirical and, therefore, contingently true or false. At the same time, since Principle Zero states that to be an empiricist means to believe that E+, the contraries of E+ are not admissible. Consequently, if it is understood in this way, in terms of the acceptance of a distinctive statement, empiricism is not tenable.

On the contrary, empiricism is tenable if it is interpreted as the adoption of a philosophical stance. A stance is composed of attitudes and commitments, both epistemic and evaluative. Disagreement about attitudes and commitments is different from disagreement about the factual content of beliefs in that the

invalid, can be set in contrast to nothing. And if what has bees said is taken literally, 'experience' denotes just this wide universe (LW1, 371)". I do not think that it makes any substantial difference to the theoretical soundness of the argument as I have presented it. The only difference that I can detect is that, in the case of philosophical experience, the conclusion of the argument is even more paradoxical. Indeed, it implies that the burden to provide a theory of everything is up to philosophy, which means to deny the cognitive autonomy of natural sciences.

former is strong enough to excise some positions as unacceptable. Take the case of science. Clearly, science is built upon the intrinsic tension between agreement and disagreement. Even though classical approaches in philosophy of science tended to be suspicious of disagreement, it is clear that peer-disagreement about scientific propositions is not only possible, but represents the standard condition of scientific research. Take now a slightly different situation, and consider the case in which a researcher refuses to submit her thesis to empirical testing and rejects with no argument all the evidence contrary to the thesis she advocates. In this case, disagreement is not to be located at the level of content, but rather at the level of methodological and heuristic standards of research. While scientists are constrained by the rule of the game they are playing to accept and handle peer-disagreement about content, they are not compelled to tolerate attitude-disagreement: they are legitimate to exclude the conclusions reached by the heterodox research from the range of what counts as science. A similar conclusion, van Fraassen remarks, can be drawn for philosophy.

Now, from a pragmatist perspective this argument is not particularly original. As is well know, it was James who first made explicit the connection between types of philosophies and kinds of attitudes or – in his own words – temperaments. The temperament of a philosopher, James remarks, "gives him a stronger bias than any of his more strictly objective premises" (James 1975, 11). Similarly as in van Fraassen's argument, James's account acknowledges a strong difference between premises and temperaments, where the latter function as 'selective filters' which provide standards of evaluation and criteria for selecting among sets of beliefs. A temperament can be seen as a standpoint from which to weight the evidence in favor of a group of theses, as well as their plausibility: a temperament "loads the evidence for him one way or the other, making for a more sentimental or a more hard-hearted view of the universe, just as this fact or that principle would" (James 1975, 11). Therefore, it is not correct to argue that stances or temperaments are fully devoid of cognitive content: as van Fraassen correctly remarks, "[s]tances do involve beliefs and are indeed inconceivable in separation from beliefs and opinions"; the point is, rather, that "a stance will involve a good deal more, will not be identifiable through the beliefs involved, and can persist through changes of belief" (van Fraassen 2002, 62).

My suggestion is to read Dewey's idea of experience as method in continuity with van Fraassen's insistence on the importance of stances for the formulation of a philosophical position. In this view, saying that experience is a method means that philosophy should reject a priori methods of investigation in favor of an empirical approach that examines the manifold life-behaviors through which, and in which, human beings interact with their environment. It is a decision taken by philosophers to prevent themselves from using certain methods of inquiry.[31] A method is not right

[31]This point is made clear by Dewey in the final chapter of his incomplete book *Modern Philosophy and Unmodern Philosophy*. In that chapter, significantly entitled *Experience as Life-Function*, Dewey lists four postulates that will guide his philosophical investigation. I quote the entire

or wrong in the same way a belief is true or false, as van Fraassen correctly points out: a method can be confirmed or disconfirmed by the consequences it leads to, but no matter how much evidence is provided in favor or against its theoretical fertility, it is always up to the researcher – in this case, the philosopher – to decide whether or not it has to be adopted. Clearly, we are prone to criticize as irrational those who would continue to apply a method that has proven itself to be unreliable. In the last analysis, this is the kernel of Peirce's idea of different and differently adequate methods for fixing beliefs. For an agent to be rational, empirical evidence should bear on the choice of which method to use in conducting inquiries. However, the choice of a standpoint or of a method is a matter of will, not of cognitive assessment of truth and falsity.

Something more can be said about the relation between attitude and evidence. I have just stressed that the adoption of a certain attitude is not compelled by the evidence available, partly because attitudes provide the criteria for valuating and weighting beliefs. Nonetheless, I have also remarked that beliefs put some constraints on the choice of attitudes: if most of the beliefs to which an attitude is committed are false, than that attitude is unreliable. Now, it is worth noticing that, by banishing some methods, the adoption of the empirical standpoint that Dewey defends also puts some constraints on the ways in which the subject-matter under investigation should be dealt with. Different methods prescribe different ways of accounting for the same set of phenomena.[32] As said, Dewey's idea of experience

---

passage, which recapitulates much of what has been said thus far: "[t]he first postulate to be set forth is that, for the purpose of the discussion, *experience* is taken as a synonym for *living* or occurrence of life-functions. The second of the postulates is that *living* and *life-functions*, as the words are here used, stand for events whose nature is most clearly and fully presented in *human* living, a fact which is equivalent in general to recognition of the socio-cultural nature of the phenomena dealt with. The third postulate is that *psychological* theory or doctrine is concerned with the analysis and description of just these phenomena, which may also, taken collectively, be named *behavior* (with "human" tacitly prefixed). The fourth postulate, underlying and giving point to the discussion as a whole, is that a correct theory of *experience* in the sense just defined is a necessary means or agency for systematic criticism of the activities, including beliefs, which at the present time constitute any existing form of living (with, be it remembered, sociocultural understood or taken for granted) and for constructive projection of the general aims and policies of such living. That critical and constructive effort of this kind constitutes *philosophy* is not so much a separate postulate as it is the focal point of the four postulates just set down" (Dewey 2012, 315–316). It is clear from this quotation that Dewey is well aware that philosophical reflection is ultimately grounded in an act of free choice which is informed and responsible to data, but is not necessarily implied by them. "The meaning of postulates," he remarks, "is to be gathered mainly from the consequences flowing from their use; these consequences also furnish the test of their value." And even more explicitly, while stressing the structural difference between philosophical and mathematical postulates, he states that "[b]ecause of the subject matter with which they are concerned they [philosophical postulates] are matters of the method of exposition of a point of view rather than of a method of demonstration" (Dewey 2012, 315).

[32]This should not look particularly controversial; it is a methodological restatement of a general principle that holds for every science. For instance, not every chemical reaction is said to have biological significance, since only those reactions that take place in cells constitute the subject-matter of biochemistry. Similarly, not every biological activity has sociological or ethnological

as method calls attention to the broad context in which human beings interact with the objects that constitute their environment. Consequently, a philosophical outlook inspired by the method of experience cannot be an object-centered approach; rather, it has to focus on the plurality of life-behaviors in which objects are experienced. In the language of contemporary philosophy, one would say that it is a practice-oriented approach.

Let's take stock and draw the consequences of what has been said so far. I have repeatedly remarked that Dewey's empirical method is to be read as a cautionary remainder of the importance of context for the comprehension of the (meaning of the) objects that are referred to by proper terms. Clearly, this holds in general: as Dewey notices, even in ordinary language the word 'experience' is used as a remainder of the context in which objects are dealt with and proper terms are employed. It follows, therefore, that attention to the broader context in which a term is applied is relevant to both science and philosophy. However, there is a remarkable difference between these two kinds of life-behaviors. In scientific research, as well as in our common-sense transactions with the environment, we do not run the risk of losing track of the context in which the activity is situated and carried out.

A life-behavior is identified and defined by its ultimate goal. Now, it is the ultimate goal of the life-behavior that sets the normative criteria for evaluating the means to reach the end, as well as the successful attainment of the latter – see below, Chap. 3, for a detailed discussion of Dewey's theory of normativity. As Dewey points out, to burn a house down in order to roast the pork living inside does not count as a successful attainment of the goal of eating roasted pork (LW13, 227). Now, when we are engaged in a scientific or common-sense activity, it is very rare that we end up forgetting or misunderstanding the goal that directs and controls our activity. It may happen, but in those cases the agent is shut out – so to say – from the activity in which she is engaged. She has lost the thread of her thoughts and actions; her activity does not make sense anymore.

Things are rather different in philosophy. It is easy, while doing philosophy, to forget that the outcome of the process of inquiry, as well as all the elements used in that process, is the result of an act of selective choice (LW1, 32). Philosophy is a reflective activity that, in order to be effective, has to select some aspects of reality as worthy of investigation, and then elaborates a set of theoretical and explanatory tools that could account for those aspects currently under investigation. As a reflective life-behavior, philosophy is on a par with science: it is grounded in idealization and abstraction. The problem with the traditional self-image of philosophy, according to Dewey, is that it is prone to forget that the conclusions to which reflection – any reflection, philosophy included – leads are *refined* products of inquiry. In so doing, philosophers misunderstand the ultimate aim of their activity,

significance, even though, after Darwin, nobody would deny that human beings are biological organisms. Simply, every perspective establishes its 'domains of objectivity' – i.e., those objects, properties and events which form the subject-matter of inquiry – where such establishment is not a priori but empirical, and it is therefore constitutively open to further revision.

which consists in providing an explanation of some phenomena, and take as their goal that of providing a complete and conclusive description of reality. Clearly, this theoretical move is not naive, being motivated by metaphilosophical, metaphysical and epistemological considerations. Nonetheless, its consequences are profound and contestable.

The main problem with that approach is that the lack of acknowledgment of the function of selection and idealization may easily lead philosophers to erase the boundaries between their theoretical concepts and the reality they purport to understand through them. Since philosophy has the tendency to "regard the results of reflection as having, in and of themselves, a reality superior to that of the material of any other mode of experience", it ends up with denying full and genuine reality to life-behaviors that are not reflective – as, for instance, much of our ordinary transactions with the environment (LW1, 26). As a consequence of that, the standard of reality becomes the one proper to intellectual experience, that is, the one established by reflective life-behaviors. From this point, it is an easy step to treat the results of reflection as independent from the reflective context in which they have been reached.

Dewey strongly insists on this point. "The commonest assumption of philosophies," he writes, "is the assumption of the identity of objects of knowledge and ultimately real objects" (LW1, 26). Such an assumption leads to consequences that fly in the face of Dewey's approach. First of all, the pluralism of life-behaviors is explicitly denied: if reality – in its proper sense – is given only in reflective life-behaviors, what is experienced in non-reflective life-behaviors is appearance at best. Secondly, by identifying reality with the objects of knowledge, the practice-oriented approach that Dewey recommends and adopts is implicitly abandoned. Indeed, the ultimate components of reality are understood as independent from any possible context of inquiry: true, they reveal themselves in and through reflection, but they do not bear any intrinsic relation to the particular kind of life-behavior in which they are experienced.[33]

[33]The expression that Dewey gives to the tendency, characteristic of traditional philosophies, to lose sight of the reflective context in which objects of knowledge are constituted is 'philosophic fallacy'. Philosophic fallacy is defined by Dewey as the "conversion of eventual functions into antecedent existence", that is, the transformation of the results of a reflective process of inquiry into autonomous and independent elements of reality (LW1, 34). As is well known, the expression 'philosophic fallacy' is a loan from William James's *Principles of Psychology*. In chapter 7, entitled *The Methods and Snares of Psychology*, James remarks that "[t]he great snare of the psychologist is the confusion of his own standpoint with that of the mental fact about which he is making his report," and that this should be acknowledged as "the 'psychologist's fallacy' par excellence" (James 2007, 196). Partially because of the language they employ, which is likely to cause confusion, psychologists are prone to substitute their theoretical explanations for the mental facts that they are investigating, thus projecting on the latter the refined distinctions that are characteristic of the former. Take, for instance, a case of knowledge – and, clearly, the example is not chosen by chance. James writes: "[t]he psychologist [...] stands outside of the mental state he speaks of. Both itself and its object are objects for him. Now when it is a *cognitive* state (percept, thought, concept, etc.), he ordinarily has no other way of naming it than as the thought, percept, etc., *of that object*. He himself, meanwhile, knowing the self-same object in *his* way, gets easily led

Against such reductionist and monist approach to the plurality of life-behaviors, the idea of experience as method brings to the fore the irreducible variety of human activities, and, in so doing, it also highlights the structural complexity of the notion of object of knowledge. We are thus led back to SIT once again. Dewey explicitly connects the idea of an empirical method with the recognition of the double-barrelledness of experience by saying that "empirical method is the only method which can do justice to [the] inclusive integrity of 'experience'", namely, the fact that subject and object are necessarily correlative notions (LW1, 19). The empirical method succeeds in doing justice to experience by calling attention to the selective choices that are made by philosophers and scientists when, in the attempt to provide an explanation of some problematic phenomenon, they adopt an object-centered perspective which focuses on the objective element of experience, leaving its subjective element on the background.[34] "Empirical method," Dewey remarks, "finds and points to the operation of choice as it does to any other event" (LW1, 34).

The recognition that the acts of choice which constitute reflective life-behaviors and, simultaneously, the objects of knowledge are far from being neutral is, therefore, the fundamental metaphilosophical justification of the validity of experience as method. Choice is an activity that – as any other overt activity that an agent may undertake – entails the modification of the environment. As Dewey formulates this insight, philosophers should realize that the object of knowledge – what he calls "the product of knowing," thus emphasizing its intrinsic connection to the activity through which it is constituted – "is the *statement* of things", and not the "*existential equivalent* of what things are 'in themselves" (LW1, 375).

Now, the understanding of the transforming power of reflection requires a consistent metaphilosophical framework and an adequate method of investigation. Here is what Dewey writes in this regard:

> What empirical method exacts of philosophy is two things: First, that refined methods and products be traced back to their origin in primary experience, in all its heterogeneity and fullness; so that the needs and problems out of which they arise and which they have to

---

to suppose that the thought, which is *of* it, knows it in the same way in which he knows it, although this is often very far from being the case. The most fictitious puzzles have been introduced into our science by this means" (James 2007, 196). Dewey's line of reasoning is an extension of James's one. Not only the psychologist, but also the philosopher is at risk of believing that the features of reality that she aims at explaining possess the same features of the thoughts or concepts (i.e., the results of reflective activity) through which she reaches the explanation of the phenomena. In so doing, the properties of a highly refined set of objects are objectified and transformed in real and ultimate qualities of reality.

[34]See, for instance, the following passage, in which Dewey refers, once again, to the function of breathing in order to formulate his views on the nature of experience: "[a]lthough breathing is in fact a function that includes both air and the operations of the lungs, we may detach the latter for study, even though we cannot separate it in fact. So while we always know, love, act for and against things, instead of experiencing ideas, emotions and mental intents, the attitudes themselves may be made a special object of attention, and thus come to form a distinctive subject-matter of reflective, although not of primary, experience" (LW1, 21).

satisfy be acknowledged. Secondly, that the secondary methods and conclusions be brought back to the things of ordinary experience, in all their coarseness and crudity, for verification (LW1, 39).

It has been remarked above that the validity of a method of inquiry consists entirely in the consequences that follow from its adoption. Dewey argues that one of the main theoretical advantages of the adoption of empirical method in philosophy is that it makes it possible to account for the relation between common-sense and refined (scientific or philosophical) objects in a clear and consistent way, by appealing to the differences between the life-behaviors in which they are experienced. In particular, adopting the empirical method amounts to taking a denotative approach that indicates the way to "go behind the refinements and elaborations of reflective experience to the gross and compulsory things of our doings, enjoyments and sufferings," that is, "to the things that force us to labor, that satisfy needs, that surprise us with beauty, that compel obedience under penalty" (LW1, 375–76).

Now, as many interpreters have pointed out, the denotation to which Dewey refers has nothing to do with denotation as conceived within the empiricist tradition. Indeed, what is pointed to, seized, and denoted are not objects, but life-behaviors or experiences – what Dewey calls primary and reflective experience. So, in order to complete the exposition of Dewey's metametaphysical assumptions, it is now time to address those notions with due detail, and show what kind of explanatory role they are meant to play in Dewey's philosophy of science.

1.5 Forms of Experience

One of Dewey's most important and yet controversial distinctions is that between primary and secondary or reflective experience.[35] It is likely that no other Deweyan notion has received so much attention – and not only by pragmatist scholars. So, for instance, because of its closeness to the later Husserlian concept of life-world, it has attracted the interest of phenomenologists[36]; in more recent times, that conceptual distinction has been translated into sociological language, and has been widely adopted and employed in the field of Science and Technology Studies.

At the very same time, that conceptual distinction is far more problematic than has usually been assumed. Many of the misunderstandings regarding the notions of primary and secondary experience are certainly due to Dewey's idiosyncratic vocabulary, which is prone to give birth to various misleading conceptual associations. To name only one remarkable example, Dewey's thesis that primary experience is non-cognitive has led many to believe that his notion is substantially identical to

[35]The bibliography on this subject is extensive. My favorite approach is Ryan (1994). But see also Eames (1964), Campbell (1995), in particular Chapter 3, Browning (1998), Garrison (2005).

[36]See, for instance, Garrison and Shargel (1988), Sukale (1976) and Kestenbaum (1977).

James's idea of baby's experience as a "great blooming, buzzing confusion" – which is clearly not what Dewey has in mind. But I also believe that some other relevant misunderstandings concerning that conceptual couple depend on Dewey's lack of clarity in distinguishing between its various components.

The goal of this final section is precisely that of disentangling those components. In particular, I aim to show that the conceptual couple primary/secondary experience should not be flattened and boiled down to two other well-known Deweyan distinctions, namely that between experience had and experience known and that between cognitive and non-cognitive (immediate) experience. Clearly, there are some similarities between those distinctions, but I argue that they play different role in Dewey's philosophy of science. While the distinctions between experience had and experience known and between cognitive and non-cognitive experience are functional through and through, the conceptual couple primary/secondary experience points to a more substantial difference between two kinds of life-behaviors.[37]

I proceed as follows. First of all, I lay out the notions of primary experience and secondary experience as formulated in the first chapter of *Experience and Nature*. Then I move on discussing the various uses that Dewey makes of those notions, with the aim of showing that they are not easily reducible one to the others. Thirdly, I introduce and elucidate the distinction between experience had and experience known, and I argue for its heuristic value. Fourthly, I highlight the difference existing between those two conceptual couples, and I claim that Dewey should have kept them more neatly separated. Finally, I will introduce and discuss the third Deweyan distinction, that between cognitive and non-cognitive (immediate) experience, and show that it is theoretically and explanatory autonomous from the other two.

Let's start with Dewey's definition of the notions of primary and secondary experience. While sketching the main features of the empirical method, Dewey writes:

> This consideration of method may suitably begin with the contrast between gross, macroscopic, crude subject-matters in primary experience and the refined, derived objects of reflection. The distinction is one between what is experienced as the result of a minimum of incidental reflection and what is experienced in consequence of continued and regulated reflective inquiry (LW1, 15).

According to that description, primary experience is characterized by a relatively little degree of what Dewey calls "incidental reflection", while reflective or secondary experience is defined in terms of the activity of inquiry. Such formulation is not problematic, at least at first glance: it simply states that primary experience

[37]The functional nature of some of those distinctions has been clearly acknowledged by Garrison. Commenting on Browning's reading of the notion of experience as starting point of inquiry, he writes: "[t]he distinction between noncognitive, immediate, unreflective experience and cognitive, mediating, and reflective experience is a functional distinction between two phases or subfunctions of experience; neither is more 'real'" (Garrison 2005, 839). The interpretation that I articulate in the following pages is a refinement of that insight.

is less cognitively refined than secondary experience. But what does it mean precisely? What does it mean that the subject-matter of primary experience is "gross, macroscopic, crude"?

As I read it, primary experience is to be understood as those set of activities, biologically determined and culturally shaped, which consist in the simplest and most basic transactions that human beings have with their fellows and the elements of their environment. Human beings have biological as well as cultural needs, which they struggle to satisfy: now, when they undertake activities of this kind, they experience things in their gross, macroscopic, crude immediacy. The latter are called by Dewey 'ordinary things', and the standard way of transacting with them is – to use Dewey's technical formula – by using and enjoying them. In primary experience things are not reflected upon for the sake of knowledge; they are directly and immediately employed for *practical* purposes.[38] As Dewey remarks in his *Logic*, "[u]se and enjoyment are the ways in which human beings are directly connected with the world about them", where use is always "for the sake of some consummation or enjoyment" (LW12, 69; see also LW16, 245).[39]

The last quotation comes from Dewey's *Logic* – and, more precisely, from Chap. 4, entitled *Common Sense and Scientific Inquiry*. It is part of the analysis that Dewey makes of the notion of common sense, with the aim of providing a satisfactory description of its main features that could pave the way for a justification of the differences between common sense and science. Now, the identification that I have, somehow implicitly, suggested between primary experience and common sense is not done by chance. Rather the contrary, I explicitly claim that the notion of primary experience should be understood as substantially coincident with that of common sense – and similarly, I argue for the theoretical interchangeability between secondary experience and science, understood in its broadest sense possible

[38]'Practical' is used by Dewey in a technical and specific sense, which is not coincident with its standard philosophical meaning. Here is what Dewey says about it: "[t]he concern of common sense knowing is 'practical' [. . .]. But *practical* in the first case is not limited to the 'utilitarian' in the sense in which that word is disparagingly used. It includes all matters of direct enjoyment that occur in the course of living because of transformation wrought by the fine arts, by friendship, by recreation, by civic duties, etc." (LW16, 253).

[39]And then he adds: "[q]uestions of food, shelter, protection, defense, etc., are questions of the use to be made of materials of the environment and of the attitudes to be taken practically towards members of the same group and to other groups taken as whole" (LW12, 69). The same point is formulated with almost the same words in *Knowing and the Known*: "[i]t is highly doubtful whether anything but matters with which actual living is directly concerned could command the attention, and control the speech usage of 'mankind,' or of an entire community. And we may also be reasonably sure that some features of life are so exigent that they impinge upon the feeling and wit of all mankind – such as need for food and means of acquiring it, the capacity of fire to give warmth and to burn, of weapons for hunting or war, and the need for common customs and rules if a group is to be kept in existence against threats from within and without" (LW16, 244–245). It seems safe to conclude from these quotations that Dewey grounds primary experience on the biological endowment of human beings, which then undergoes a process of cultural refinement and articulation that eventuates in the plurality of social forms of life. I have tried to elaborate on this Deweyan insight in Gronda (2015a).

as including all those activities that are not reducible to the practical concerns of human beings.[40]

When the distinction between primary and secondary experience is read in this way, the relations between the two notions should appear less mysterious. As said, primary experience is the set of activities concerned with the satisfaction of practical purposes. Secondary experience is understood as that particular kind of activities or transactions with the (social or natural) environment, which are characterized by what Dewey calls "the concern of inquiry *as inquiry*" (LW16, 254). In the passage from *Experience and Nature* that I have quoted above, such distinctive feature of secondary experience is formulated by saying that, in this case, things are "experienced in consequence of continued and regulated reflective inquiry". Secondary experience is, therefore, a specific mode of practice, whose essence consists in "the advancement of *knowing* apart from concern with *other* practical affairs" (LW16, 255). It comes as no surprise that Dewey treats philosophy as a form of secondary experience, which differs from scientific activity narrowly conceived on the ground of its relation to the subject-matter of common sense. More on this in a moment.

Recent philosophy of science has devoted little attention to investigating the relations between science and common sense. On the contrary, the clarification of those relations is, for Dewey, a task of the utmost importance, to such an extent that he dedicates to the analysis of this topic not only the chapter in his *Logic* that I have mentioned above, but also one in *Knowing and the Known* – significantly, the only chapter of the book written by Dewey alone, without the collaboration of Bentley.

[40]It would not be difficult to provide arguments in support of this thesis. However, since I believe that this point is of relatively little interest to the present discussion, I rest content with listing a few passages relating to the notions of common sense and science – all drawn from *Knowing and the Known* – which show that the identification of primary experience with common sense is corroborated by textual evidence. (1) "The words 'occupied, engaged, concerned, busied,' etc., repay consideration in connection with the distinctive subjectmatter of common sense. *Matter* is one of the and-so-forth expressions"; (2) "The words 'concern,' 'affair,'. 'care,' 'matter,' 'thing,' etc., fuse in indissoluble unity senses which when discriminated are called *emotional, intellectual, practical*[...]. The supremacy of subjectmatters of concern, etc., over distinctions usually made in psychology and philosophy, cannot be denied by anyone who attends to the fact"; (3) "The other consideration is even more significant. What has been completely divided in philosophical discourse into man *and* the world, inner *and* outer, self *and* not-self, subject *and* object, individual *and* social, private *and* public, etc., are in actuality parties in life transactions" (LW16, 246–248). The notion of common sense is used here to play the same role – and to stress the very same points – that it is played by the notion of primary experience in the first chapter of *Experience and Nature*. See also what Dewey says in the chapter of the *Logic* entitled *Common Sense and Scientific Inquiry*: "(1) Scientific subject-matter and procedures grow out of the direct problems and methods of common sense, of practical use and enjoyments and (2) react into the latter in a way that enormously refines, expands, and liberates the contents and the agencies at the disposal of common sense[...]. When scientific subject-matter is seen to bear genetic and functional relation to the subject-matter of common sense, these problems disappear" (LW12, 71–72). Here it is the notion of scientific subject-matter that seems to be coincident with that of secondary experience, as employed in *Experience and Nature*.

But why is Dewey so concerned with that issue? Or, put in other terms, what are the functions that he believes the two conceptual couple can carry out?

Dewey's account of the relation between primary and secondary experience – or between common sense and science – is highly articulated; I focus, therefore, only on three aspects that are more relevant for my purposes.[41] First of all, the distinction between primary and secondary experience accounts for the genesis of science as a distinctive mode of practice which is guided by what Dewey once calls "the *ideal* of scientific knowing in the *moral* sense of the word" (LW16, 253). Science originates from common sense as a consequence of a process of transformation and refinement of the tools of inquiry as well as of the attitudes and activities of the communities of agents. In this sense, science is an extremely sophisticated cultural product which is the result of various choices that have been made by some – but not all, and in this sense are far from being necessary – human societies. Dewey's genealogical account of Western philosophy – recall that philosophy is reflective experience just as science is – is precisely meant to show the contingency of the problems and concerns with which a discipline deals. As there are no perennial problems of philosophy – which means that there are no problems which human beings must *necessarily* address – so it is not necessary for human beings to develop and adopt a scientific attitude: there may well be environmental conditions that do not foster or encourage the formation of the scientific spirit.[42]

Secondly, the distinction between primary and secondary experience, modeled on that between common sense and science, enables Dewey to explain the theoretical difference existing between philosophy and science, narrowly conceived. As repeatedly remarked, philosophy and science are both forms of secondary experience: they are reflective activities which have as their goal the explanation of some problematic features of the things constituting the subject-matter of primary experience. There is a remarkable difference between the two, however. While science stems from and goes back to the subject-matter of primary experience, thus succeeding in providing a clarification of the latter, philosophy tends to forget its roots in primary experience as well as the explanatory task that it is expected to carry out. It is worth quoting the following passage at length to illustrate Dewey's views on that issue:

> The charge that is brought against the non-empirical method of philosophizing is not that it depends upon theorizing, but that it fails to use refined, secondary products as a path pointing and leading back to something in primary experience. The resulting failure is three-fold. First, there is no verification, no effort even to test and check. What is even worse, secondly, is that the things of ordinary experience do not get enlargement and enrichment of meaning as they do when approached through the medium of scientific principles and reasonings. This lack of function reacts, in the third place, back upon the philosophic subject-matter in itself. Not tested by being employed to see what it leads to in ordinary

[41]The fourth aspect – which concerns the difference in structure between scientific and common-sense *concepts* – will be discussed in the next chapter, Sect. 2.6.

[42]Dewey's long sojourn in China (1919–1920) made him suspect that there might even be cultural conditions that hinder the adoption of the experimental method. I have discussed the philosophical implications of Dewey's sojourn in China in Gronda (2015c, 2017).

> experience and what new meanings it contributes, this subject-matter becomes arbitrary, aloof what is called "abstract" when that word is used in a bad sense to designate something which exclusively occupies a realm of its own without contact with the things of ordinary experience (LW1, 16–17).

The separation of secondary from primary experience has remarkable, obnoxious consequences. Clearly, the role of reflection is that of suspending overt activities, thus making it possible for the agent to construct a mode of response apt to handle new environmental stimuli. Science and philosophy – as highly refined forms of reflection – purify and perfect this function of reflection, constituting themselves as activities guided by the ideal of knowing. The problem with philosophy, therefore, is not that of loosening the connection to primary experience; rather, it is that of misapprehending the nature and function of such loosening. In so doing, philosophy ends up postulating theoretical entities which do not lead back to primary experience. This implies, on the one hand, that the meaning of ordinary things is not enlarged and enriched by the objects of secondary experience; on the other hand, that the entities postulated by philosophy are taken to be completely autonomous and independent from ordinary things. As a consequence of that, philosophers are prone to believe that the objects that they postulate are real "in and of themselves" (LW1, 17).[43] But, it is only when that assumption is endorsed that the philosophical problems of understanding the relation between the subject-matters of primary and secondary experience arise. Those are not problems that have to be solved; indeed, as all the problems that stem from the adoption of a non-empirical method of reflection, they are "blocks to inquiry, blind alleys", "puzzles rather than problems" (LW1, 17). Rather, they have to be dissolved by rejecting the very method which gives rise to them.[44]

The latter remarks introduce the third use that Dewey makes of the conceptual couple primary/secondary experience. In the long passage quoted above, Dewey

[43] A word of caution is in order here. One may advance an argument of this sort. Let's assume that the theoretical entities postulated by philosophical investigations are not real "in and of themselves", as Dewey suggests to do. Let's also assume that philosophy and science are structurally identical, since they are both secondary experience. Accordingly, it follows that the objects postulated by our best science too are not real "in and of themselves". If all these assumptions are true – and I believe they are – then, so the argument goes, the conclusion can be drawn that Dewey is a scientific anti-realist. No matter how sound the argument may seem, the conclusion is rash: Dewey strongly emphasizes that he is a scientific realist, and that he believes that scientific objects are real. See Chap. 5 for an in-depth analysis of Dewey's scientific realism.

[44] Compared with genuine scientific problems, the artificiality of traditional philosophical problems is apparent. Dewey is clear about the difference between them. See, for instance, the following passage: "[t]he refined objects of reflection in the natural sciences, however, never end by rendering the subject-matter from which they are derived a problem; rather, when used to describe a path by which some goal in primary experience is designated or denoted, they solve perplexities to which that crude material gives rise but which it cannot resolve of itself. They become means of control, of enlarged use and enjoyment of ordinary things. They may generate new problems, but these are problems of the same sort, to be dealt with by further use of the same methods of inquiry and experimentation. The problems to which empirical method gives rise afford, in a word, opportunities for more investigations yielding fruit in new and enriched experiences" (LW1, 17).

makes it explicit that one of the main differences between scientific and philosophical entities consists in their capacity to undergo verification. Scientific objects lead back to experience by indicating the ways in which they can be verified. What Dewey has in mind here is the empirical confirmation of Einstein's theory of general relativity offered in 1919 by the observation of a solar eclipse in South Africa (LW1, 15). However, this insight can be extended beyond that particular, paradigmatic case. In order to articulate it, Dewey uses a metaphor: "the vine of pendant theory," he writes, "is attached at both ends to the pillars of observed subject-matter" (LW1, 11).

Now, the problem is that of understanding what Dewey means with that statement. At first glance, its significance is clear: it means that scientific hypotheses have to be tested experimentally, by producing observations that could confirm them. However, it is important not to misunderstand Dewey on this point: as I read him, his is not a profession of verificationism, at least according to the standard interpretation of that concept. Dewey's position is much more nuanced.

Clearly, Dewey is committed to the thesis that scientific hypotheses can be verified through observation. This thesis may well appear naive to our eyes, but a couple of cautionary remarks are worth giving so as to avoid jumping to too hasty conclusions. First of all, a historical consideration: Dewey writes *Experience and Nature* – from what all those passages are drawn – in 1925, almost ten years before the publication of Popper's *The Logic of Scientific Discovery*, to name only one remarkable attempt to criticize and revise our intuitions about the idea of verification. No surprise, therefore, that Dewey's views on that matter are not as refined as we would like them to be.

The other consideration is more substantial, and revolves around Dewey's idea of observation. When speaking of observation, a contemporary reader is naturally led to think of the logical empiricist account of it. According to the latter, the language of a theory is constituted by three kinds of terms – logical, theoretical, and observational – and observational language is defined as that language which contains only observational and logical terms. Consequently, one of the most urgent issues that philosophy of science is expected to address is that of elaborating a set of criteria for distinguishing between theoretical and observational terms.

Dewey's appeal to observation as a means for verification is only superficially similar to the logical empiricist line of thinking. Dewey does not aim to provide criteria for distinguishing between observational and theoretical terms, or between observational and theoretical languages; rather, he is concerned with stressing the different functions performed by two different kinds of *activity*. Recall Dewey's discussion of the empirical confirmation of Einstein's theory of relativity; analyzing the philosophical import of that event, he writes:

> An Einstein working by highly elaborate methods of reflection, calculates theoretically certain results in the deflection of light by the presence of the sun. A technically equipped expedition is sent to South Africa so that by means of experiencing a thing – an eclipse – in crude, primary, experience, observations can be secured to compare with, and test the theory implied in, the calculated result (LW1, 15).

In light of this passage, it seems safe to argue that Dewey shifts attention from linguistic reports to the very activity of observation that brings about those reports. More precisely, I read him as saying that the activity of observation has necessarily to do with some kind of manipulation of the things of ordinary experience. Clearly, Dewey does not deny that the members of the expedition that travel to South Africa to take pictures of the eclipse were technically equipped; nor is he restrictive about the use of theoretical terms in the formulation of the observational reports. He does not take a position on this issue. What he is interested in is the fact that observation is to be understood as an activity that *necessarily* belongs to primary experience.

Textual evidence can be easily found in support of an interpretation along this line. I mention only two passages that I find particularly significant and compelling. First of all, Dewey explicitly states that the experienced material – the pillars of observed subject-matter to which the vine of pendant theory is attached at both ends, to use Dewey's analogy – "is the same for the scientific man and the man in the street". Obviously, he qualifies that statement by noticing that the man in the street needs special preparation in order to follow and understand the intervening theory. Nonetheless, it remains true, he holds, that "stars, rocks, trees, and creeping things are the same material for both" (LW1, 11).

Secondly, and most interestingly, consider what Dewey says about the Michelson-Morley experiment – which was originally performed with the aim of detecting the existence of aether, and whose negative result played some role in Einstein's formulation of his theory of relativity. It is worth quoting the passage in full:

> when the Michelson-Morley experiment disclosed, *as a matter of gross experience*, facts which did not agree with the results of accepted physical laws, physicists did not think for a moment of denying the validity of what was found in *that experience*, even though it rendered questionable an elaborate intellectual apparatus and system. The coincidence of the bands of the interferometer was accepted at its face value in spite of its incompatibility with Newtonian physics [...]. Not for a moment did [scientific inquirers] think of explaining away the features of *an object in gross experience* because it was not in logical harmony with theory as philosophers have so often done [...]. In short, the material of refined scientific method is continuous with that of the actual world as it is concretely experienced (LW1, 38; italics mine).

Unless one is ready to attribute to Dewey the thesis that the content of the observational reports is matter of ordinary experience – a thesis that is both textually and theoretically untenable, since it is apparent that the construction of observational reports (the coincidence of the bands of the interferometer) requires a high degree of scientific and technical preparation – it seems clear that the only possible way to account for that statement is by acknowledging that Dewey aims to stress that scientific observation depends on the same *activities* that constitute primary or ordinary experience. Put in other terms, scientific observation has necessarily to do with the manipulation of middle-size objects that are encountered in perception: no matter how refined its conceptual content may be, what is perceptually discriminated always belongs – *at some level* – to the subject-matter of primary experience. The facts disclosed by the Michelson-Morley experiment are matter of gross experience in the sense that human beings can *see* that the bands of the interferometer coincide. And this is made possible, in the last analysis, by the biological endowment common to all human beings.

So, to recapitulate, the third use of the distinction between primary and secondary experience, centered around the notion of verification, enables Dewey to achieve two different goals. On the one hand, it enables him to explain the role that observation and evidence play in scientific inquiry. Observed facts are stubborn: it is their stubbornness that makes them effective as tests for scientific theories. Clearly, observations can be variously interpreted – as will be shown in Chap. 4, the acknowledgment of the theoretical malleability of observations is one of the distinctive features of Dewey's theory of evidence. Nonetheless, their malleability is not indefinite. Dewey argues that the limits to the theoretical malleability of observations depend on the fact that, at some point, scientific theorizing gives way to primary experience. Or, as he formulates this insight, scientific hypothesis ends in the making of "new observations and experiments among the things of raw experience" (LW1, 15). There might be no reasons for that, but it is a fact that primary experience is the ground of intersubjective agreement. Scientific theorizing requires training and preparation to be understood, while seeing the coincidence of the bands of the interferometer is open to everyone.

On the other hand, such a distinction makes it possible for Dewey to avoid relativistic conclusions stemming from a too relaxed interpretation of the idea of the malleability of observations. Recall Dewey's metaphor of the vine of pendant theory which is attached at both ends to the pillars of observed subject-matter. As said, the point that Dewey wants to make with that metaphor is that the observed subject-matter is the same for the man of science and the man of the street. This means that there is always a common ground to which both the man of science and the man of the street have access through their life-behaviors. As will be shown in Chap. 2, this insight has relevant semantic consequences that Dewey articulates in his pragmatist account of concepts. But it also implies – though Dewey does not draw such a conclusion – that men of science, no matter how different their scientific reasonings may be, do always have the chance to understand each other by going back to that common ground provided by common sense. In the last analysis, common sense is named so because of its capacity to constitute a common world for human beings. The primacy that Dewey gives to common sense over science has to do with the inescapability of primary experience.

The metaphor of the vine of pendant theory, however, reminds of another well-known Deweyan distinction, that between experience had and experience known.[45] Traditionally, that distinction has been formulated as follows: while some activities are concerned with the acquisition of knowledge, some others are concerned with the enjoyment, use, ad consummation of things. The former are instances of experience known; the latter of experience had.

---

[45] See, for instance, Hildebrand (2003, 145) or Anderson (2006, 132).

Dewey is not very clear about that distinction.[46] Nonetheless, it is not difficult to glimpse its theoretical import when it is read in light of Dewey's theory of inquiry. I will discuss the structure of inquiry in Chap. 3, so I rest content here with outlining some of its constitutive features. According to the pragmatist conception of knowledge, which Dewey undoubtedly endorses, inquiry originates from the perception of a problem and ends when that problem is solved. Recall how Peirce describes doubt and belief in terms of the relation that one bears to the other:

> Doubt is an uneasy and dissatisfied state from which we struggle to free ourselves and pass into the state of belief; while the latter is a calm and satisfactory state which we do not wish to avoid, or to change to a belief in anything else (Peirce 1986, 247).

Doubt or inquiry arises as a response to a state of dissatisfaction, and, when appropriately conducted, eventuates in a satisfactory state of calm.[47] Clearly, the achieved state of calm will not last indefinitely: once a state of belief is reached, the agent will act in conformity to the content of that belief, thus putting it to the test of practice. Action leads to new problems which lead to new doubts and inquiries.

Within this framework of analysis, the couple experience had/experience known is functional to distinguish and demarcate two different – temporally different – phases of inquiry. In this sense, the pillars to which the vine of pendant theory is attached are to be understood as those phases of experience in which the agent is not engaged in an act of conscious reflection, i.e. phases which precede and follow the activity of inquiry.

I do not want to argue that a reading along this line of Dewey's metaphor is illegitimate. My point is more modest: I simply want to highlight that that distinction cannot be equated with the one between primary and secondary experience. Indeed, if they were identical, it would follow that primary experience should be devoid of any cognitive trait – which is textually false. Consider, for instance, the following passage drawn from *Knowing and the Known*:

> Doing and knowing are both involved in common sense and science – involved so intimately as to be necessary conditions of their existence. Nor does the difference between common sense and science consist in the fact that knowing is the important consideration in science but not in common sense. It consists of the position occupied by each member in relation to the other. In the concerns of common sense knowing is as necessary, as important, as

[46]What is usually considered Dewey's most detailed analysis of this conceptual couple runs as follows: "But experienced situations come about in two ways and are of two distinct types. Some take place with only a minimum of regulation, with little foresight, preparation and intent. Others occur because, in part, of the prior occurrence of intelligent action. Both kinds are *had*; they are undergone, enjoyed or suffered. The first are not known; they are not understood; they are dispensations of fortune or providence. The second have, as they are experienced, meanings that present the funded outcome of operations that substitute definite continuity for experienced discontinuity and for the fragmentary quality due to isolation" (LW4,194).

[47]For the sake of simplicity I here use the notions of doubt and inquiry as substantially interchangeable, though I am aware they are by no means identical. Please see below, Sect. 3.1, for a discussion of their conceptual differences. At this stage and at the present level of analysis, however, distinguishing between the two would have yielded no further relevant information.

> in those of science. But knowing there is for the sake of agenda, the what and the how of which have to be studied and to be learned – in short, known in order that the necessary affairs of everyday life be carried on (LW16, 252; see also LW12, 71).

It is clear from this quotation that the notion of common sense – which I take to be substantially identical with that of primary experience – is meant to encompass those inquiries that are directed to solving common-sense problems, that is, those problems that arise from the use of ordinary things for practical purposes. This reading is confirmed, among the other things, by Dewey's insistence on the structural identity of common-sense and scientific inquiries.

The difference between the two conceptual couples can be, therefore, expressed in the following terms. The distinction between primary and secondary experience is substantial, so to say, as it is intended to distinguish between two different kinds of life-behaviors – or better said, between two different classes of life-behaviors; on the contrary, the distinction between experience had and experience known is functional, and refers to different ways of handling a certain subject-matter – be it of common sense or science. No surprise that the two conceptual couples can hardly be boiled down to a common denominator.

Before concluding this chapter, however, it is worth discussing another relevant conceptual distinction that Dewey employs in his writings – namely, that between cognitive and non-cognitive (or immediate) experience. That distinction is originally formulated in the series of early articles that Dewey devotes to outlining his immediate empiricism, and then is further articulated in *Experience and Nature*. The rationale behind that distinction is that, even in the case of reflective experience, experience presents some non-cognitive features that are non-emendable. In this sense, that distinction expresses the classical pragmatist insight that it is not possible to doubt everything: for a doubt to occur, there must be other elements that are kept exempted from doubt and directly acted upon.

The following quotation is particularly pertinent. It is drawn from an important footnote of *Experience and Nature*, in which Dewey clarifies what his criticism of intellectualism amounts to be. Dewey's goal is to argue for the theoretical inconsistency of the idea of what he calls "the alleged all-inclusiveness of cognitive experience". He writes:

> It is not denied that any experienced subject-matter whatever may *become* an object of reflection and cognitive inspection. But the emphasis is upon "become"; the cognitive never is all-inclusive: that is, when the material of a prior non-cognitive experience is the object of knowledge, it and the act of knowing are themselves included within a new and wider non-cognitive experience – and *this* situation can never be transcended (LW1, 30).

Much could be said about this passage. I focus on two possible interpretations of the entanglement of cognitive and non-cognitive elements in an experience that are relevant to Dewey's philosophy of science. On the one hand, the rejection of the all-inclusiveness of cognitive is grounded in the acknowledgment of the fact that, as any other life-behavior, a cognitive activity has a focus and a background, and that the focus is made possible by that pattern of anticipations, expectations and memories that constitute its background. The use of the conceptual couple cognitive/non-cognitive experience points, therefore, to the fact that any act of cognition is guided by a purpose which is not explicitly formulated, but rather immediately felt. If interpreted in this way, that distinction is intrinsically connected to Dewey's idea of tertiary quality – and his aesthetic account of normativity – which I will deal with in Chap. 3.

But that conceptual couple can also be interpreted in a slightly different way. Indeed, it can be read as an attempt on Dewey's part to formulate the distinction between know-how and know-that – or, differently put, between tacit and explicit knowledge.[48] If read this way, Dewey's point is to stress that cognitive experience is possible only because an agent can rely on habits of behavior that she does not question, and which provide the solid ground on which her conscious, cognitive activity rests. Those unquestioned habits of behavior are the non-cognitive elements that make cognitive experience possible.

Take, for instance, a chemist who is trying to synthesize a new molecule in laboratory. Clearly, her attention is focused on those aspects of the experimental setting which are new to her and relevant to the success of her activity: for instance, she carefully observes and controls the multiple transformations of the reactants. However, all the body of chemical knowledge that she employs in the experiment, as well as all the practical skills that enable her to manipulate the reactants and to "see" what is going on there, are not consciously expressed or formulated by the agent. She does not state them as general rules of behavior, neither does she need to grasp them as propositions in order to be able to carry out her laboratory work. Rather, those skills work *tacitly*; they are multi-layered strata of reliable habits of behavior that can be linguistically expressed if needed, but whose effectiveness as guides of action depends entirely on their capacity to satisfactory anticipate the environmental responses in a way that does not call out reflection or inquiry.

Now, the important point to note for present purposes is that the distinction between cognitive and non-cognitive experience cuts across the other two distinctions previously analyzed. Indeed, the entanglement of cognitive and non-cognitive is present both in primary and secondary experience, and is as important in the case of practical life-behaviors as it is in the case of activities directed to acquiring knowledge. Similarly, both reflective and non-reflective experience present a focus and a background, and cannot be properly distinguished on that basis only. This means that the conceptual couple cognitive/non-cognitive experience is genuine,

---

[48]As is well known, these two distinctions can be traced back, respectively, to Gilbert Ryle and Michael Polanyi. Though they are now treated as identical, these distinctions were introduced for different purposes. Ryle appeals to the notion of know-how in order to counteract the regress argument to the effect that any comprehension of a proposition presupposes a prior comprehension of another proposition (Ryle 1945). On the contrary, Polanyi uses the notion of tacit knowledge to highlight the importance of scientific tradition and practice in science (Polanyi 2009). In recent time, much has been written on those notions, as a consequence of the practical and contextualist turn both in epistemology and philosophy of language. For a broad-brush overview of how the concepts of know-how and tacit knowledge have been declined and developed within the pragmatist tradition, see the special issue on pragmatism and common sense in the *European Journal of Pragmatism and American Philosophy*, and in particular Hetherington (2017).

since it cannot be traced back to, or derived from, the other two: it has theoretical and heuristic autonomy.

To recapitulate. The goal of this section was that of highlighting the variety of uses of the notion of experience that Dewey makes in his works, and to offer a plausible and usable taxonomy of the different conceptual couples through which he articulates the content of that notion.[49] It has been shown that, in some cases, the concept of experience is used to convey logical ideas; in some other cases, it has a metaphilosophical value; in still others, it is employed as replacement of epistemological notions, whose theoretical legitimacy Dewey openly and explicitly denies. In the following chapters these distinctions will be repeatedly called into question, discussed and further investigated. In so doing, the theoretical complexity of Dewey's practice-oriented approach to philosophy of science will emerge in its fundamental aspects.

[49]In conclusion, a word of caution may be useful. It is worth reminding that I do not hold that the three conceptual couples that I have here sketched can be found clearly distinguished in Dewey's writings. Rather the contrary, Dewey often uses those notions in quite a loose way. Take, for instance, the following passage from *Experience and Nature*: after affirming that intellectualism "goes contrary to the facts of what is primarily experienced", Dewey states that "things are objects to be treated, used, acted upon and with, enjoyed and endured, even more than things to be known". And then he concisely summarizes this insight by saying that those things "are things *had* before they are things cognized" (LW1, 28). Here Dewey seems to somehow equate the notion of primary experience with that of experience had. I have quoted this passage as a proof of the fact that it is easy to find textual evidence supporting a different account of these notions. So, to name only one remarkable case, in his introduction to Dewey Fesmire identifies primary and immediate experience – see (Fesmire 2014, 62). All this notwithstanding, however, I think that it would be better to keep those conceptual couples separated since, in so doing, the explanatory power of Dewey's conceptual apparatus gets enhanced. In any case, I am aware that my theoretical reconstruction of these Deweyan notions is strongly prescriptive.

Chapter 2
Language

Abstract This chapter is devoted to the analysis of Dewey's philosophy of language, which is central for the understanding of the semantic account of inquiry I formulate in Chaps. 3 and 4. It also strengthens and refines the conclusions reached in Chap. 1, thus providing a more fine-grained analysis of the structure of life-behaviors. Dewey's theory of inquiry is grounded in the distinction that Dewey draws between signs and symbols, and between significance and meaning. I locate the difference between significance and meaning in the kind of response that signs and symbols solicit, and, in so doing, I explore the relevance of Dewey's semantic externalism and anti-representationalism.

Keywords Language · Metaphilosophy · Activity · Sign · Symbol · Meaning · Significance · Agreement in practice · Pragmatist account of concepts · Operationalism · Holism · Articulation · Scientific and common-sense concepts

2.1 Introduction

The previous chapter has been devoted to outlining a comprehensive framework in which to situate Dewey's notion of experience. Several aspects have been discussed: it has been pointed out that Dewey defines experience as an infinity-word; that he understands experience as a method rather than as a substance; that he argues that experience always refers to the broader context in which things are encountered; and that he thinks of experience as a particular type of activity performed by an organism with a highly complex biological endowment. Finally, it has been highlighted that, according to Dewey, experience is intrinsically related to the notion of meaning: experience is the *locus* in which and through which natural events become meaningful and, in so doing, acquire a new level of complexity by participating in our transactions with the world.

Dewey holds that the realization of these potentialities of nature is achieved when human beings start using language. Thanks to language, human beings can behave in new and more articulated ways: they stop responding in a mechanical way to physical stimuli, and starts answering to the relevant features of environment –

R. Gronda, *Dewey's Philosophy of Science*, Synthese Library 421,
https://doi.org/10.1007/978-3-030-37562-1_2

relevant to them – with a purpose in view, in the context of a shared course of activity. Meaning is, therefore, ultimately grounded in language; it is the latter which makes it possible for biological organisms – which, as natural entities, are subjected to direct physical interactions, and, as biological entities, are capable of converting physical things into materials sustaining life – to respond to physical and biological stimuli as meaningful entities.

Surprisingly, however, Dewey's philosophy of language has been almost completely neglected by interpreters, both within the pragmatist and the analytic tradition. Apart from a recent symposium edited by David Hildebrand on the *European Journal of Pragmatism and American Philosophy*, very few articles have dealt with the subject, and, to my knowledge, no book-length study has been ever dedicated to it.[1] The goal of this chapter is to clarify Dewey's conception of language, and, in so doing, to prepare the ground for the *semantic* interpretation of Dewey's logical theory that I will articulate in Chap. 4. That interpretation hinges upon the twin distinctions between sign and symbol, and significance and meaning, and is therefore understandable only in the context of Dewey's philosophy of language.

Before starting off with a detailed reconstruction of Dewey's philosophy of language, a terminological remark is needed. Up to now, I have used 'meaning' in a broad and rather loose way to refer to the general phenomenon of *being meaningful*. Now, in *this* sense of the word everything can have meaning – a word as well as a thing. This use is not arbitrary: in *Experience and Nature*, for instance, Dewey often uses 'meaning' in this way, as when he writes that human beings respond "to things in their meanings", or when he defines objects as "things with meanings" (LW1, 278). This use of the word is loose but not slack. The insight that Dewey wants to convey with those statements is the idea that human beings respond to things in ways that are structurally different from the mechanical and physiological interactions of natural bodies with the environment. Human beings are capable of perceiving – and 'perceiving' and 'perception' are technical notions of Dewey's philosophy, as will be seen later in this chapter – the possible consequences that may ensue from the modification of some existential conditions. These consequences are then projected on what is currently present, and act as a factor that influence and modify the course of activity that results from that act of deliberation.

However, somewhere between the publication of *Experience and Nature* and the writing of the *Logic* Dewey changes his mind about the nature of meaning. While in his earlier texts meaning is understood in opposition to existence – consider, for

[1] Among the contributions to that symposium see, in particular, Johnson (2014) and Dreon (2014). To my knowledge, the first attempt to shed light on Dewey's philosophy of language is Mesthene (1959). Quite important – and I will deal with it at length in the following pages – is Black (1962). In recent times, a more balanced reconstruction of Dewey's views on language has been advanced by Pratt (1997) and Midtgarden (2008). See also Ryan (2011), chapter 6. A very interesting approach is Boersema (2008), which purports to show the contribution of pragmatism to contemporary analytic philosophy of language. On this issue, see also Fesmire (2014, 110–112).

instance, the title of the article *Existence and Meaning*[2] – in *Logic* the notion of meaning is defined in relation to that of symbol, and in contrast to the concept of signification. 'Meaning' becomes, therefore, a purely *linguistic* notion, which refers to the way in which non-natural signs operate. As a consequence of this refinement, which is conceptual as well as terminological, Dewey's account of meaning as formulated in the *Logic* and in his latest works – as, for instance, in the essays written in preparation for, or immediately after the publication of, *Knowing and the Known* – cannot be traced back, immediately and without further clarification, to the analyses that can be found in his earlier texts.

As is clear, such terminological shift entails numerous interpretative problems, starting with the fact that it becomes much harder to provide an unified account of Dewey's mature philosophy because of its internal articulation. Consequently, understanding the relation between the idea of experience as the *locus* of meaning, as outlined in the previous chapter, and the linguistic account of meaning, as will be reconstructed in this chapter, is far from straightforward. I recommend a two-step approach. First of all, let's start with distinguishing the two notions on a terminological level: I call the broader notion of meaning – the one outlined in Dewey's earlier texts, *Experience and Nature* included – 'Meaning_B', while I refer to the linguistic notion of meaning simply as 'meaning'.[3] In so doing, confusions and easy misunderstandings should be avoided. Secondly, once the two notions are clearly demarcated one from the other, I suggest looking for the philosophical reasons that lie behind Dewey's decision to refine the notion of Meaning_B, and to substitute it with the more refined concepts of meaning and significance.

In order to do that, it is useful to stress an important point, which will be discussed at length in the following pages, and which is central to my understanding of Dewey's logical theory. As should be clear from what has been said thus far, I hold that Dewey's earlier notion of Meaning_B is much less refined than the

---

[2]I do not want to dwell into this issue here, since a detailed discussion of this point will lead us astray, but it is worth reminding that Dewey draws the conceptual couple meaning/existence from Bradley – in particular, from his *Principles of Logic*. Dewey's early article *Knowledge as Idealization* shows rather clearly why Dewey found that distinction so useful. Indeed, that distinction enabled him to understand knowledge in terms of the processes of thought through which brute facts of sensation were transformed into elements endowed with meanings. As Dewey puts it, in order to understand "the aspect of meaning or significance" one has to turn his attention to the "content of the idea as opposed to its existence" (EW1, 177). Leaving aside the historiographical technicalities, my concern is to point out that conceptual couple existence/meaning is a theoretical tool that originates in Dewey's early philosophy, and is then preserved and employed – at least, this is how I read Dewey's philosophical development regarding this issue – until he fully realizes the importance of the function of language for the process of meaning-constitution. I have analyzed Dewey's terminological debt to Bradley in Gronda (2012, 86–101).

[3]Since in the present chapter I will deal almost exclusively with meaning in this latter sense, I prefer to make the reading more manageable without introducing unnecessary notations. So, though in the following pages I will also use meaning in the standard, non-Deweyan sense, I have decided not to employ a specific expression to mark this usage of the word, in the belief that it should be clear from the context when the term is used in this way.

linguistic notions of meaning and significance. Consequently, I also believe that *Logic* represents a significant improvement in the expressive resources of Dewey's philosophy. Nonetheless, a clarification is in order here. Despite its lack of clarity, the notion of Meaning_B has the merit of highlighting the continuity that, according to Dewey, exists between the meaning of symbols and the significance of natural signs – the very same kind of (semantic) continuity that is grasped by what I have called, in the previous chapter, the Semantic Identity Thesis. Though meaning and significance are functionally different, they are not structurally distinct; rather, they are to be understood as two different and yet intercommunicating forms of human behavior – or, better said, of human ways of answering to environmental stimuli. Indeed, no conceptual or linguistic distinction is so fine-grained as to be impossible for it to get embodied in a direct course of activity: the content of a linguistic symbol is open and on its way to become the content of a natural sign and, consequently, of an object. There is no metaphysical and ontological break between mind and world, symbol and sign, meaning and significance: this because, within Dewey's practically-oriented framework, both are kinds of operation. Dewey's use of the notion of Meaning_B – as well as of its cognate notion of experience – aims to underline precisely the possibility of such passage from the linguistic level to the level of external reality.

With that said, let's turn attention to Dewey's mature philosophy of language, remembering that the analyses contained in this chapter are meant to integrate the conclusions reached in the previous one. The chapter is structured as follows. The first section is dedicated to outlining some essential features of Dewey's approach: by stressing its main points of difference with what I have called – rather imprecisely, I admit – the Fregean tradition, my goal is to avoid some possible misunderstandings that are likely to arise from an uncharitable reading of Dewey's texts. The second section introduces the fundamental distinction between sign and meaning, as formulated by Dewey in his *Logic: Theory of Inquiry*. In the third section I will discuss the particular version of externalism that I believe can be carved out from Dewey's scattered remarks on the nature of language. The fourth section deals with Dewey's pragmatist theory of concepts, while in the fifth and final section I investigate and assess Dewey's distinction between common-sense and scientific concepts.

2.2 Language as Activity: A Shift of Paradigm

I have said above that Dewey's philosophy of language has not received adequate attention in the literature, and I have remarked that such dearth of attention is surprising considered the central role that language plays in Dewey's philosophy, and, in particular, in his logical theory. It has to be acknowledged, however, that, though theoretically puzzling, it is not difficult to find some historical and contingent reasons for why that neglect has occurred. On the one hand, mainly because of Rorty's provocative re-interpretation of the pragmatist heritage, much of the contemporary debate has been centered on the experience vs. language dichotomy. In order to stress the difference between classical and analytic pragmatism, many pragmatist scholars have insisted on the irreducibility of experience to language, and have focused on the former to the detriment of the latter. This choice – which was (at least partially) motivated by polemical purposes – had unfortunate consequences in that it shut down a highly fruitful line of investigation.[4]

Since I have discussed at some length that point in the previous chapter, I will not dwell further on it here. On the contrary, I will focus on another possible cause of the neglect of Dewey's philosophy of language among contemporary interpreters. Put it roughly, I would say that the idiosyncratic character of Dewey's account of language made it difficult for analytic philosophers to understand and appreciate the fruitfulness and originality of his approach. The most explicit example of such a difficulty of comprehension is Max Black's article *Dewey's Philosophy of Language* (1962).

In that article, Black charges Dewey with lack of clarity as well as with numerous conceptual confusions. I deal with only two of them here, which are particular relevant for the purpose of clarifying the differences between the two approaches. First of all, Black criticizes Dewey for failing to distinguish between meanings of the words and meanings of the sentences, ending up subsuming them under the same logical category (Black 1962, 518). From Black's point of view, Dewey does not properly understand the principle of compositionality.[5] According to the classical formulation of the principle, the meaning of a complex expression is determined by the meanings of its components *plus* the rules of semantic composition of language. Clearly, some version of the principle of compositionality is necessary for any theory of language in order to account for the capacity that human beings display to understand (and generate) sentences that they have never heard (and uttered) before. The problem with Dewey's approach is that it does not seem to understand the function played by the rules of language in the construction of a sentence. In this sense, Black argues that "a statement is not properly to be regarded as a complex word – any more than a house is properly to be regarded as a complex brick – or a brick as a rudimentary house". The two notions belong to two distinct logical categories. For this reason, "any account of meaning that is presented as applying indifferently to statements and to their components is bound to be at least systematically ambiguous" (Black 1962, 519).

The other confusion which Black chastens Dewey for concerns the latter's account of meaning. Black reads Dewey as endorsing the view that the meaning of a symbol can be identified with its "*nonverbal counterpart*" – a counterpart that can be somehow independently designated, and which exists "independently of the symbol whose meaning it is alleged to be" (Black 1962, 519). No matter how ingenuous such an account may be, Black argues, that is but another way of formulating, in terms more palatable to contemporary philosophical tastes, the standard empiricist

[4]For a lucid analysis of this predicament, see Johnson (2014).

[5]See, on this point, (Pratt 1997, 855).

thesis – which Black attributes to Locke – according to which the meanings of the words are made up of the thoughts that the speaker has in her head, which are clothed in words only for the sake of communication. The point is that if meaning is understood as something which is independent of language – for instance, as a method of action or as a pattern of regular behavior – then the relation between words and meanings would become external and purely arbitrary. Language would therefore be not necessary for the possibility of meaning – a conclusion which not only would be philosophically problematic, but would also fly in the face of almost everything Dewey has written on this topic.

I do not want either to assess the soundness of Black's criticisms or to debate whether or not Black is a careful Deweyan reader. Others have already done this task in a satisfactory manner.[6] My aim with reporting Black's remarks was simply to call attention to the sense of strangeness that one of the very few analytic philosophers who took Dewey's philosophy of language seriously had in trying to make sense of his reflections on that issue. More precisely, my ultimate aim is to advocate interpretative prudence, so as to be careful not to jump to too hasty conclusions. I concede that much of what Dewey has written about language and meaning does not meet the standards of clarity and precision set by contemporary analytic philosophy of language. However, once this fact has been acknowledged – and *has* to be acknowledged – I think one should refrain from concluding that Dewey's philosophy of language is fatally flawed. Similarly, one should refrain from concluding that Black completely misses the point in ascribing the views that he criticizes to Dewey. Actually, things are much more complicated than that. The goal of this section it to suggest that a third option is available.

Clearly, in order to make that option seem plausible, it has to be shown that the objections directed against Dewey's approach are not conclusive. For the sake of simplicity, I will take into account here Black's criticism that Dewey has failed to distinguish between words and sentences, thus preventing himself from realizing that the meaning of a sentence cannot be boiled down to the meanings of its components.[7] The thesis that I would like to defend is that the real point of disagreement between Dewey and the Fregean tradition is metaphilosophical rather than properly philosophical.

First of all, let's start with acknowledging that Black is right: actually, Dewey lacks the distinction between words and sentences. Such difference is clear and evident for those who, like Black, work within a Fregean framework, but it is not for Dewey, who explicitly rejects it. But how is it possible? How is it possible that Dewey fails to appreciate that distinction? One plausible answer is simply that he is wrong, and that he does not understand the semantic role played by syntactic structures in the composition of meaning. I am not willing to say that an answer along these lines cannot be upheld – it definitely can be. However, that amounts to

[6]As, for instance, Pratt (1997) and Midtgarden (2008).

[7]I could have decided to analyze the other criticism advanced by Black, and the conclusion would have been the same. In any case, I will tackle Black's reductionist criticism below, in Sect. 2.4.

saying that Dewey misses an important point that is central to the Fregean tradition; from which it is an easy step to conclude that his philosophy of language is flawed. But that is nothing but a wholesale rejection of his approach: which is clearly a legitimate position, but one which any interpreter should be ready to endorse only as an *extrema ratio*. For my part, I think a different – and more charitable – reading can be advanced, which tries to locate that shortcoming in the broader context of his thought, with the aim of clarifying the philosophical reasons that lead Dewey to endorse his incomplete and partially unsatisfactory conception of meaning.

Here is my suggestion, which elaborates on a point made by Torjus Midtgarden in his article *Dewey's Philosophy of Language*. In that text, while discussing Dewey's conception of language as the cultural matrix of inquiry, Midtgarden stresses that "a synchronic study of language as an 'inclusive code' (LW12, 55) should be anchored in a broader account of social behaviour, communicative behaviour in particular" (Midtgarden 2008, 260). Leaving aside for now the broader implications of this remark for the analysis of Dewey's theory of inquiry, I read Midtgarden as saying that in order to grasp Dewey's conception of language, one should focus on his understanding of language as a mode of behavior. Following this insight, I hold that the ultimate philosophical reason why Dewey fails to acknowledge – or decides not to acknowledge – the difference between meaning of words and meaning of sentences is that the goal of his philosophy of language is less that of understanding the structure of linguistic propositions, or sentences, than that of clarifying what kinds of behavior are made possible by language. And within this practice-oriented paradigm of analysis, Dewey believes that the demand of clearly distinguishing between meaning of words and meaning of sentences has to be sacrificed in favor of the theoretical necessity of not jeopardizing the primacy and unity of linguistic *behavior*.[8]

Let's look closer at Dewey's argument. Its grounding assumption is Dewey's wariness towards abstraction.[9] Philosophical abstraction tends to overshadow the complexity of the thing that it should instead explain, and it easily ends up substituting the refined results of the process of analysis for the thing itself. Such tendency is particularly evident in the case of philosophical reflection about language: in this case, indeed, it is easy to hypostatize the linguistic medium, and to consider it as a sort of 'third' element that mediates, in different ways, between mind and world. In so doing, one is led to forget that language is primarily a way of behaving, a tool – actually, Dewey states that it is "the tool of tools" (LW1, 146) – which is used to achieve some specific goal in a specific situation, and has to

[8] A word of caution is in order here. With the formula "primacy and unity of linguistic behavior" I do not want to convey the idea that linguistic behavior should be treated as something monolithic. It is clear that linguistic behavior is internally diversified, and covers the totality of human transactions with the world. Rather, the idea of primacy of linguistic behavior purports to highlight the fact that linguistic behavior is to be understood as a whole, and that, taken as a whole, it has a substantial and explanatory primacy over its components. In this sense, the notion of linguistic behavior is structurally identical to that of experience – see above Sect. 1.3.

[9] On this topic, please see Winther (2014).

be primarily investigated in light of the function it carries out in those activities. It follows, therefore, that no distinction between the linguistic elements employed in a certain life-behavior can be formulated once for all, without paying attention to the particular conditions in which they occur.

This is precisely what Dewey objects to logical positivists in his *Logic*. In the context of a critical overview of the advantages and drawbacks of logical positivism's choice to use "sentences" and "words" instead of the more traditional "propositions" and "terms", he writes:

> A [...] serious objection is that without careful statement, the new terminology does not discriminate between language that is adapted to the purposes of communication (what Locke called 'civil' language) and language that is determined solely by prior inquiries related to the purposes of inquiry – the latter alone being logical in import. This serious difficulty cannot be overcome by considering sentences and words in isolation, for the distinction depends upon an intent which can be adjudged only by means of context (LW12, 284).

It is important to realize what Dewey does *not* want to assert in that passage.[10] Indeed, one might read Dewey as saying that a word may function as a sentence in some contexts of utterance.[11] However, despite his reference to the contextual

---

[10]For the sake of completeness, it has to be reminded that there is another argument that Dewey offers in support of his refusal of adopting the strong distinction between words and sentences advocated by logical positivists. Since this argument has an exclusively historical interest, I limit myself to outline its main points here. Dewey's starting point is the acknowledgment that the distinction between meanings of the words and syntactical relations – though perfectly legitimate in itself – is often used to support the much less uncontroversial one between form and matter. The latter distinction is controversial not in the sense that no distinction between form and matter should be admitted – Dewey is clearly not an eliminativist about that conceptual couple. As he writes, "there is no question that logical theory must distinguish between form and matter". However, "the necessity for the distinction does not decide whether they are or are not independent of each other"; in particular, it does not decide "[w]hether they are or are not, for example, intrinsically related to each other in logical subject-matter and distinguishable only in theoretical analysis" (LW12, 285). The problem is, therefore, not that of denying the theoretical legitimacy of that distinction, but rather that of understanding in which terms it should be accounted for. The issue at stake is what counts as a good philosophical explanation.

[11]It is likely that we are reminded here of the famous Wittgensteinian analysis of the very same distinction in *Philosophical Investigation*. "But what about this: is the call "Slab!" in example (2) a sentence or a word? a If a word, surely it has not the same meaning as the likesounding word of our ordinary language, for in §2 it is a call. But if a sentence, it is surely not the elliptical sentence "Slab!" of our language. – As far as the first question goes, you can call "Slab!" a word and also a sentence; perhaps it could aptly be called a 'degenerate sentence' (as one speaks of a degenerate hyperbola); in fact it is our 'elliptical'sentence. a But that is surely only a shortened form of the sentence "Bring me a slab", and there is no such sentence in example (2). But why shouldn't I conversely have called the sentence "Bring me a slab" a lengthening of the sentence "Slab!"? a Because anyone who calls out "Slab!" really means "Bring me a slab". a But how do you do this: how do you mean that while saying "Slab!"? Do you say the unshortened sentence to yourself? And why should I translate the call "Slab!" into a different expression in order to say |9| what someone means by it? And if they mean the same thing, why shouldn't I say, "When he says 'Slab!' he means 'Slab!' "? Again, why shouldn't you be able to mean "Slab!", if you can mean "Bring me the slab!"? – But when I call out "Slab!", then what I want is that he should bring me

sensitivity of the distinction between sentences and words, Dewey's aim is not to argue for the importance of context for the determination of the content of the manifold elements of language. True, semantic externalism is one distinctive feature of Dewey's conception of language, so a reading along that line would not end up attributing to him a thesis that he does not actually endorse. Since I will discuss Dewey's externalism below in Sect. 2.4, I do not dwell on it here any further. Nonetheless, I think that in the passage quoted above Dewey wants to make a slightly different point, which concerns the variety of life-behaviors in which human beings can be engaged, and the philosophical resources that are needed in order to account for them in a satisfactory way.

As has been highlighted in the previous chapter, one of the most distinctive Deweyan ideas is that language functions differently in different life-behaviors, and that the categories with which to deal with those functions should vary in the same way. In particular, when it is used for the purpose of communication language works differently from how it works when it is used in the context of inquiry. That insight can be formulated in multifarious ways. I like how Barrotta puts it: within the process of inquiry, he remarks, terms or concepts are in search for their meaning; on the contrary, in those contexts in which language is used for communication meanings are already settled, and no problem with their conceptual content arises (Barrotta 2018, 21–27).[12]

Dewey's point is that assuming the distinction between meaning of the words and meaning of sentences is likely to lead to serious logical fallacies when applied to the analysis of inquiries. Take a paradigmatic case of communication. In this case, Dewey remarks, a speaker tries to say something to a hearer: for instance, she wants to communicate the fact that the dog is lost. Now, the grammatical subject, "the dog", is taken to be shared by both the speaker and the hearer, and its content is held fixed. It is the hinge on which the whole proposition turns. The content of the predicate "is lost" is, on the contrary, what has to be communicated: it can

a slab! — Certainly, but does 'wanting this' consist in thinking in some form or other a different sentence from the one you utter?" (Wittgenstein 2009, §19).

[12]This trait of Dewey's thought has usually been either misrepresented or neglected by his critics. So, for instance, it has often been argued that Dewey's pragmatist account of inquiry amounts to a wholesale rejection of any form of correspondentism in favor of some sort of anything-goes approach to truth and knowledge. I think this is not correct: I believe that Dewey would be ready to admit that when everything is in order – as it is in the case when language is used for the purposes of communication – language actually refers to objects in the world, and true beliefs correspond to state of affairs. Dewey's point – at least, as I read him – is that these experiences are not amenable to logical investigation. Nor should they be analyzed from an epistemological perspective. They are complex forms of linguistic behavior, characterized by their own intrinsic normativity. These linguistic behaviors can be profitably investigated from various points of view, but are philosophically rather uninteresting. True, it is important to provide a philosophical account of how it is possible that facts and ideas, things and concepts, *come to have* the same content. One of the main goals of his *Logic* is precisely that of providing an account of this "correspondence". But that is a genetic and temporal account which purports to shed light on a distinctive feature of that particular life-behavior that Dewey labels "inquiry". It cannot be extended indefinitely to other types of life-behavior.

be assumed that that information is in possession of the speaker, and has to be transmitted to the hearer. In this cases, the distinction between meaning of the words and meaning of sentences is unproblematic, and can be relied upon for the purpose of understanding how communication works.

Now, take a genuine case of inquiry. In this case, the meanings of the words are, by definition, not settled: terms are here in search of their meaning. So, there is not a fixed content on which the inquirer can rely. Indeed, if there were, meanings would be established, and inquiry would be closed. When it comes down to investigating the structure of reflective life-behaviors, Dewey argues, the distinction between meaning of the words and meaning of sentences is therefore worse than useless. It is harmful since, in Dewey's own terms, "if the logical theory of the subject-predicate is taken over from grammatical structure", then it is highly likely that the content of the logical subject will be conceived of as "something already completely given independently of inquiry and of the need of inquiry" (LW12, 285). This means that the process of inquiry cannot be interpreted in terms of the attribution of a predicate to an already existent subject. If such approach is adopted, the differences between those life-behaviors in which language is used for communication and those in which it is used for the purposes of inquiry are blurred, and the specificities of the latter are thus overlooked. Inquiry is a process of predication as well as of "subjection": the meanings of the subject and the predicate of the judgment are reciprocally constructed (LW12, 135). Accordingly, the complexity of inquiry requires different and more refined conceptual tools.

I will deal with Dewey's theory of inquiry in Chap. 4. In that context, I will further investigate the theoretical apparatus that Dewey elaborates to account for the distinctive features of inquiry, and I will better clarify not only the reasons of his rejection of the distinction between words and sentences, but also his positive account of the relations between terms and propositions in the process of construction of judgment. For the moment, I am content with highlighting that his rebuttal of that distinction can be explained not only in terms of his alleged lack of understanding of how language works, but also on the basis of a precise meta-philosophical agenda which has to do with the theoretical resources that Dewey thinks are needed to account for the plurality of uses of language. As said, the same conclusion could have been arrived at in other ways, so I do not want to put too much weight on this particular Deweyan argument. For present purposes, it is sufficient to recognize that the heuristic and explanatory difference between Dewey's practice-oriented framework of analysis and the traditional analytic paradigm is a sign of a more radical difference between the two approaches that has to be cashed out in metaphilosophical terms. More precisely, it concerns the preliminary decision as to whether adopt a practical standpoint which considers activity as the fundamental category to understand and explain language. Or, put in other terms, it concerns the decision as to whether consider language primarily as a mode of activity or as a product: Dewey opts for the first alternative, standard analytic philosophy of language for the second.

Before moving on to examine in detail Dewey's philosophy of language, however, I would like to call attention to an important consequence of his approach, which enables to explain an interesting – and, at that time, rather puzzling – Deweyan statement, namely, his rejection of the tripartition between syntax, semantics and pragmatics championed, amongst others, by Charles Morris and Carnap.

Now, the reason why metaphilosophical assumptions are philosophically relevant is because they bear specific philosophical consequences. As a matter of fact, Dewey's metaphilosophical insistence on the theoretical primacy of the categories of activity and linguistic behavior paves the way for his adoption of the philosophical thesis of the primacy of the linguistic element over its components. The fundamental insight is that linguistic sign too is a whole whose components are distinguishable only in and through theoretical analysis; consequently, any attempt to understand it as the result of the combination of allegedly simpler elements is bound to fail.

It is with this in mind that in his article *Peirce's Theory of Linguistic Signs, Thought, and Meaning*, Dewey openly criticizes "the account of meaning as pragmatic advanced by Morris on the alleged authority of Peirce" (LW15, 141). Dewey's starting point is the rejection of Morris's dualistic approach to linguistic signs. As is well known, Peirce defines sign as "anything which is so determined by something else, called its Object, and so determines an effect upon a person, which effect I call its interpretant, that the later is thereby mediately determined by the former" (EP2, 478). Dewey charges Morris with parcellizing out this triadic relation, thus obtaining three dyadic dimensions:

> The dyadic 'relation of signs to that to which they are applicable' is called the *semantic* dimension; 'the relation of signs to one another' is called the *syntactical* dimension; while 'the relation of signs to interpreters' is called the *pragmatic* dimension. It is further added that in their semantic dimension, signs designate and/or denote; in their syntactic dimension, they implicate; in their pragmatic dimension, they express (LW15, 142–143).

From Dewey's perspective, this kind of approach is problematic in that it purports to "'solve' problems by a distribution of subject-matters into different compartments", while a genuine solution requires that the various compartments of linguistic sign are seen as integrated into a whole (LW15, 143).[13] Morris's account of language implies that linguistic sign operates somehow mechanically along the

[13]It is worth remarking that Dewey does not challenge the theoretical legitimacy of these three different approaches to the study of language. His position is much more nuanced, and has to do with the risk of a metaphysical hypostatization of distinctions made for a definite explanatory purpose. See, on this specific issue, what he writes in the unpublished manuscript *What is to be a Linguistic Sign or Name* (1945): "[w]hen due precautions are taken, it is possible to distinguish three aspects of Naming – being Named in such a way as to render each capable of specialized treatment. The result is (a) study of things established as sings, whether popularly or scientifically, apart from what they name – apart, that is, from what the are *as* signs [...]. (b) It is obviously possible to study what is designated or 'signi*fied*', apart from specific study of signs as such. When this selective emphasis is converted into systematic (or 'philosophic') discussion it results in theories which imply that 'Objects' are one independent matter, something else – thoughts,. or speakers or whatever, another, and names a third intervening thing. (c) Process of *application* involved in the inclusive behavioral operations may be selected for special study" (LW16, 308–309).

three different dimensions that allegedly constitute it: the linguistic sign refers to an object in the world; then, independently of its semantic capacity to denotate, it relates to other signs; and finally, independently both of its semantic and syntactic properties, it brings about some effect on its interpreter.

Dewey's insistence on the Peircean notion of Thirdness is crucial in this regard. As Peirce gets never tired of reminding, Thirdness cannot be reduced to a combination of dyadic relations. So, for instance, the relation of giving – which is genuinely triadic – cannot be reduced to a couple of dyadic relations such as "a put b down" and "c picked b up" (Misak 2004, 21). Similarly, linguistic sign – being a case of Thirdness – cannot be understood by being treated as a sum of those dyadic relations that go under the labels of syntax, semantics and pragmatics. Indeed, trying to account for the intrinsic structure of linguistic sign in this way means to completely misunderstand its nature. Indeed, it amounts to denying the structural and categorial difference between linguistic signs – which Dewey calls "symbols" – and indexical signs. Dewey is clear on this point:

> To connect *things* with *indexical* signs is, in Peirce's position, a way of *denying* that they are connected with *linguistic* signs, with words; or anything he calls a *symbol*. For an indexical sign is a case of what Peirce calls *Secondness*, while a linguistic sign is a case of *Thirdness*. (LW15, 146).

As will be discussed at length in the next section, the distinction between natural signs and symbols is central to Dewey's naturalistic theory of language. No surprise, therefore, that he projects it on Peirce's philosophy, thus interpreting the latter's conceptions of Secondness and Thirdness in light of his own philosophical assumptions. In any case, more fundamentally than that, Dewey's criticism centers on the conceptual confusion that Morris introduces in the theory of signs, which stems from blurring the distinction between the different ways in which indexical and linguistic signs operate. While indexical signs refer directly to some object in the world, to which they are causally connected, linguistic signs or symbols "have of themselves no reference to 'things' " (LW15, 146).[14]

Morris's approach is, therefore, contrary to both the letter and the spirit of Peirce's thought. It is also contrary to what Dewey believes a sound naturalistic account of language should be. From his perspective, a naturalistic account is characterized by two constraints. First of all, it must not employ metaphysical notions; secondly, it must show how the *explanandum* can be accommodated within a naturalistic framework, clarifying the ways in which it interacts and emerges from simpler environmental transactions. According to Dewey, Morris's dyadic account of language is unsatisfactory on both accounts. On the one hand, it is heir of the classical epistemological tradition, which is in its turn dependent on

[14]Incidentally, this is one of the ways in which Dewey formulates the anti-representationalist insight that, in the case of linguistic sign, representation is not the key explanatory notion. But there is something more in Dewey's argument: he seems to imply that semantic representationalism is intrinsically connected to an unsatisfactory theory of language which relies on some metaphilosophically troublesome assumptions. I will repeatedly come back to this point in this chapter.

the metaphysical distinction between form and matter, subject and object, mind and world. On the other hand, it is anti-naturalistic since in order to explain some of the features of linguistic signs it endows words "with the same miraculous power formerly attributed to mind or to an idea as a go-between", and, in so doing, blocks the road to any possible genetic explanation of the manifold functions performed by language (LW15, 146).

According to this theoretical reconstruction, the explanation of the nature of linguistic signs is intrinsically interwoven with an anti-naturalistic account of their capacity to refer to an object in the world. Within a naturalistic framework, on the contrary, the capacity of linguistic signs to refer to objects should be explained in a completely different fashion, by paying attention to the ways in which "linguistic signs interlock with indexical signs" – or, differently put, to the "interlocking of linguistic with non linguistic *modes of behavior*" (LW15, 148 and 149; italics mine).

The shift from what may be called a representationalist account of reference to an account based on the different functions carried out by different modes of behavior has two remarkable philosophical consequences. On the one hand, it paves the way for the formulation of a naturalistic solution to the problem of the origin of language. Indeed, it thus becomes possible to provide a sound genetic account of the relations between linguistic and non linguistic modes of behaviors. As Dewey remarks, Peirce's approach can be reformulated with the following words – a reformulation which, clearly, expresses more faithfully Dewey's views than Peirce's:

> While he [Peirce] does not use the following mode of speech it is, I believe, faithful to his position to say that in the course of cosmic or natural evolution, linguistic behavior supervenes on other more immediate so to say, physiological modes of behavior, and that in supervening it also intervenes in the course of the latter, so that through this mediation regularity, continuity, generality become properties of the course of events, so that they are raised to the plane of reasonableness (LW15, 149).

On the other hand, it sets the ground for an experimental investigation of the linguistic phenomenon. Indeed, the metaphilosophical shift of attention towards modes of behavior makes it possible for Dewey to formulate a philosophical project centered around the idea of seeing "language, with all its speakings and writings, as man-himself-in-action-dealing-with-things" (LW16, 50). In this perspective, "[b]eing a name or *linguistic* sign is a matter of the performance of behavioral operation", whose semantic, syntactic and pragmatic properties arise and can only be understood in the context of the broader activity in which it is applied (LW16, 307–308).

Clearly, those assumptions put constraints on the kinds of explanation that can be considered satisfactory within Dewey's practically-oriented framework of analysis. With this in mind, let's turn to discussing the main features of Dewey's philosophy of language, starting from his naturalistic account of linguistic signs – in Dewey's terminology, symbols. Then I will move to analyze Dewey's externalism, and finally I will deal with his particular version of anti-representationalism.

2.3 Significance and Meaning: Dewey's Theory of Signs and Symbols

Human beings are natural organisms which respond to a variety of stimuli from the environment. Some of them trigger a mechanical response. Think, for instance, at knee-jerk reactions: your knee is hit, and it immediately jerks. In these cases, there is no reflection occurring between the stimulus and the response. Furthermore, the stimulus does not convey any information: it simply happens, and then a reaction takes place. Since I will discuss this point at some length in the next chapter, I won't dwell on this kind of stimuli-response relationship any further here.

Some other stimuli, on the contrary, trigger a more complex response. Take, for instance, a smoke that you perceive in a corner of the room in which you are currently sitting. The perception of the smoke is likely to prompt you to leave the room or to carefully inspect the spot from which the smoke is coming. The stimulus conveys, therefore, some piece of information: from the existence of the smoke you can infer the existence of a fire that is not currently part of your perceptual field.

Finally, suppose that you have followed the smoke and you have detected the fire in the room. You start running out of the place while you bump into a friend of yours who is unawarely entering the room. You then try to warn her by saying "Be careful, there is a fire in the room!". In this case, the stimulus is provided by a series of sounds uttered by the speaker in a certain context, which prompt the hearer to leave the room. It is important to remark here that the hearer's action is not caused by the tone in which the words are uttered, but by the content of the utterance.

What is the difference between these three kinds of response? Dewey's answer is that the first one counts as a physical reaction to the environment, and, consequently, can be satisfactorily accounted for in purely physiological terms. Instances of this kind are "the act of jumping when a sudden noise is heard, withdrawing the hand when something hot is touched, blinking in the presence of a sudden increase of light, animal-like basking in sunshine". As Dewey remarks, "[s]uch reactions are on the biological plane" (LW12, 48). These interactions with the environment are quite rare, since the responses that human beings give to external stimuli are usually mediated by their being included in a larger activity. Human beings have purposes, and the 'semantic importance' of a stimulus depends on the function it plays within a specific course of activity.[15]

The second kind of response – which has been exemplified above by the perception of smoke – is more complex precisely because the stimulus is not taken at its face value, so to say, but is used as a sign for further actions directed toward

[15] I use the expression 'semantic importance' to refer to the fact that something acts as a sign for an agent within a specific course of action. As will be clear, I use importance instead of other, more perspicuous terms such as meaning and significance because the latter are technical terms in Dewey's philosophy. It is Dewey himself who suggests this terminological choice: "importance is the generic term; significance and meaning are the specific ways in which the issue of importance has to be dealt with" (LW16, 331).

achieving a specific goal. The smoke is, therefore, not simply noticed; it is *perceived* as something which points toward a certain direction, and that can lead to the realization of an intended goal. Dewey is clear on this point:

> To *perceive* in the sense of *observing* is identical, positively and negatively, with observing in the sense of paying heed to what is observed as *directive* or *sign* in further actions. When things seen, heard, touched, smelled, are *observed*, noted, heeded, they are treated as serving a specific function or office. The office in question is that serving an end-in-*view*, or purpose which is entertained; hence it involves a need to be *tested* by its meeting or failing to fulfill the end for the sake of which it was used as a *sign*; in consequence, a fact which involves concern for additional observation, namely, the consequences actually occurring as distinct from those wanted and held in view (LW16, 322).

When an organism is capable of using things as signs, its transactions with the environment radically change. It stops being pushed and pulled by external stimuli, and starts responding to them with an eye to their possible, future consequences. Such a capacity is a precondition of meaning in that it makes intelligent behavior possible. It also introduces a temporal element in conduct since present stimuli are responded in light of the purpose that the agent wants to achieve in the future.

The third kind of response implies a higher degree of complexity.[16] In this case, symbols are embodied in physical signs, without which they could not exist and operate. However, symbols do not operate in virtue of the specific qualities possessed by their material support. Indeed, the same material support, put in different context, issues in different outcomes: symbols are – to use traditional terminology – conventional or artificial. If, while running out the room, I had told the person who was entering "Attenzione, c'è un incendio lì dentro", an English hearer would not have grasped what was going.

The latter remark is clearly a platitude, but Dewey states it for a specific purpose, which is that of neatly demarcating the way in which symbols function from the way in which natural signs do. Usually, the distinction is drawn by stressing that in order for symbols to function as signs, the participants in the conversation need to share some kind of background knowledge, be it some set of rules or some sort of practical knowledge. Obviously, Dewey acknowledges this point – I will discuss his

[16] In *Experience and Nature* the same insight is formulated in terms of the notion of experience. Dewey's argument runs as follows: "Some consequences of the interaction of things concern us; the consequences are not merely physical; they enter finally into human action and destiny. Fire burns and the burning is of moment. It enters experience; it is fascinating to watch swirling flames; it is important to avoid its dangers and to utilize its beneficial potencies. When we name an event, calling it fire, we speak proleptically; we do not name an immediate event; that is impossible. We employ a term of discourse; we invoke a meaning, namely, the potential consequences of the existence. The ultimate meaning of the noise made by the traffic officer is the total consequent system of social behavior, in which individuals are subjected, by means of noise, to social coordination; its proximate meaning is a coordination of the movements of persons and vehicles in the neighborhood and directly affected. Similarly the ultimate meaning, or essence, denominated fire, is the consequences of certain natural events within the scheme of human activities, in the experience of social intercourse, the hearth and domestic altar, shared comfort, working of metals, rapid transit, and other such affairs" (LW1, 149–150).

notion of agreement in conjoint action in the next section; nonetheless, he also thinks that that is not the main point of difference between symbols and natural signs. For reasons that will be made clear in the following exposition, he rather focuses on the fact that symbols are not tied down to their original conditions of application. Contrary to natural signs, whose referential structure is fixed, symbols can easily have their meaning extended or changed as a consequence of their being applied in new and different contexts. Which means that their semantic function is made more complex by their capacity to participate in different ways to different life-behaviors that eventuate in different outcomes.

Here is Dewey's argument:

> A "natural sign", by description, is something that exists in an actual spatial-temporal context. Smoke, as a thing having certain observed qualities, is a sign of fire only when the thing exists and is observed. Its representative capacity, taken by itself, is highly restricted, for it exists only under limited conditions. The situation is very different when the *meaning* "smoke" is embodied in an existence, like a sound or a mark on the paper. The actual quality found in existence is then subordinate to a representative office. Not only can the sound be produced practically at will, so that we do not have to wait for the occurrence of the object; but, what is more important, the meaning when embodied in an indifferent or neutral existence is *liberated* with respect to its representative function. It is no longer tied down (LW12, 58).

Symbols and natural signs function differently. As will be remembered, it is for this reason that Dewey so neatly rejects Morris's conception of semantics as a sort of fixed relation between a word and its denotatum. Such a conception stems from a 'categorial' confusion between signs and symbols, which ends up with projecting on the latter some of the properties of the former. In order to prevent this confusion from arising, Dewey uses two different terms to refer to the two ways in which natural and linguistic signs have "meaning".

Properly speaking, a natural sign does not have meaning; it has significance. Its capacity to refer to something different than itself – which is what significance amounts to – depends on a natural relation which connects one element to the other. Because of the natural relation holding between them, the first element can be taken as evidence of the existence of the second element, as a consequence of which the latter, which is not currently present, can be reliably inferred from the observation of the former. Take the case of smoke as natural sign of fire: smoke is evidence of the existence of fire, and the causal relations between the two support the inference from the actual observation of smoke to the possible future observation of fire. In Dewey's terminology, sign, significance, evidence, inference and involvement (or connection) are all related notions.

On the contrary, a symbol has meaning. Its capacity to refer to something different from itself depends on an agreement reached in action: it is only within a certain context of activity that a certain element comes to be connected with some other. Since meaning is not a natural relation, words or symbols do no provide evidence of any existence whatsoever. From my uttering the word "there is smoke in the room" the existence of fire cannot be inferred. Symbols do not operate at this level; rather, that utterance implies, at the level of discourse, that I cannot state that "there is no smoke in the room", on pain of inconsistency. Similarly, the symbol "water" implies – but it does not infer – the symbol "H_2O" in certain contexts of discourse and activity. Symbol, meaning, implication and relation do all belong to the same categorial space. Finally, reference is the term used by Dewey to name the 'relation' which symbols sustain to their objects.

Dewey's distinction between significance and meaning is straightforward, and highlights a remarkable difference in the way in which natural and artificial sign function. However, one should not be misled by the way in which Dewey formulates that distinction. Indeed, contrary to what it may seem at first glance, Dewey is not committed to the view that symbols and natural signs are neatly separated. If he endorsed that view, his whole theory of inquiry would be intrinsically contradictory.

I will deal with that issue at length in Chap. 4, which is entirely devoted to presenting my semantic reading of Dewey's theory of inquiry. For now, it is sufficient to note that Dewey acknowledges a continuity between signs and symbols, which makes it possible for the content of the latter – its meaning – to progressively become part of the relation of significance of the former. Here is an example of what Dewey has in mind. While introducing the topic of the relations between science and common sense in the fourth chapter of the *Logic*, Dewey observes that "rocks as minerals *signify* something more in a group that has learned to work iron than is *signified* either to sheep and tigers or to a pastoral or agricultural group" (LW12, 66; italics mine). In this passage, the emphasis is put on the semantic contribution of technological activities, but the relevant consequence of such emphasis is that the naturality of the relation of significance should not be understood in substantial terms, as entailing some sort of immutability due to its allegedly natural origin. The naturality of natural signs – as opposed to the artificiality of symbols – is rather to be understood as stemming from two interrelated features of the relation of significance: (1) the concrete, real presence of the first element of the relation, from which the one which is not currently under observation can be inferred; (2) the fact that it is possible to directly interact with, and operate on, the elements connected by the relation of significance. The last remark is central to Dewey's theory of inquiry: it is precisely because the manipulation of the relation among symbols does not entail any real effect in the world – which may turn out to be irreversible and dangerous for the agent – that taking the way of ideas has proven to be effective and evolutionarily convenient.

A couple of remarks are useful here in order to dispel some possible misunderstandings. Take, once again, the relation between the observation of smoke and the existence of fire. It seems here that the relation of significance should be understood as being natural in a stronger sense than the one outlined above. Indeed, since that relation is somehow parasitic on the causal relation that holds between the two events, its naturality seems to follow directly from the naturality of the causal relation. Unless one is anti-realist about causation – a thesis that Dewey does not endorse, at least in my view – there should be no problem in endorsing a more substantial conception of the naturality of the relation of significance. So, why is Dewey trying to weaken the naturality of the relation of significance? And, how is it possible to satisfactorily argue for that view, once it is shown that naturality can be understood in an ontological fashion?

This line of reasoning has some plausibility. However, it tends to radically downplay the difference between a natural relation like that of causality and a semantic relation like that of significance. The difference that thus goes unnoticed is that something is a sign only because, and insofar as, it is part of a course of activity in which the first element of the relation of significance is used to infer the existence of the second one. It does not make any sense to call something a sign of something else apart of, and independently of, its being an element of a behavior. Then, though it is true that the existential relation between smoke and fire is ontologically independent, the smoke is a natural sign of fire not in itself, but only in connection with an organism capable of observing the former, retaining it in its memory, and finally using it to infer the latter.

That line of reasoning misses, therefore, the metaphilosophical primacy that Dewey attributes to the categories of activity and life-behavior. To a certain extent, it makes a mistake similar to the one mentioned above, for which Dewey blames Morris and logical empiricists. In both cases, an attempt is made to conceive of the nature of signs as something which can be specified and determined independently of the function they play in human life-behaviors. This approach has several limits. On a more technical level, it conceals the various interactions that continually arise between meaning and significance within a course of action – and of which the case of technological activity mentioned above is an example. On a more general level, it prevents from seeing them as two moments within a broader *naturalistic* continuum, which has to be understood in terms of the kind of activities that are made possible by the biological endowment of the organism.

It has been remarked that a certain element becomes a sign only at the presence of an organism which is able to employ it in a course of activity. In order for natural signs to exist at all, the behavior of the organism must reach a certain degree of complexity, which ultimately depends on its biological constitution. I will deal with this issue at length in the next chapter, in the context of the analysis of Dewey's idea of the pre-logical conditions of inquiry, since it is relevant to the comprehension of his logical theory. For now, the point that I would like to stress is that the more complex the behavior, the more connected are the elements of the environment by the relations of *significance*. Smoke can be a sign of fire only for an organism that is capable of observation, memory, and some kind of reasoning. However, human organisms are complex enough to undertake *linguistic* behavior, which puts them in a position to broaden the scope and quality of their transactions with the environment. Language is a game changer in that it enables the agent not only to more easily retain in memory the fact of the existence of the observed element of the relation of significance, but also to detect and discover new connections.

Let's delve deeper into this insight. Dewey explicitly states that, in the case of human organisms, it is constitutively impossible to settle the question of historical priority of significance-connections over meaning-relations. Far from being two independent and autonomous forms of "relations", they are two elements that cooperate in linguistic behavior. As Dewey writes, "ability to treat things as signs would not go far did not symbols enable us to mark and retain just the qualities of things which are the ground of inference". And then he clarifies how this cooperation of signs and symbols applies to the case of smoke-fire too:

> Without, for example, words and symbols that discriminate and hold on to the experienced qualities of sight and smell that constitute a thing "smoke", thereby enabling to serve as a sign of fire, we might react to the qualities in question in an animal-like fashion and performs activities appropriate to them. But no inference could be made that was not blind and blundering (LW12, 61).

Leave aside for the moment Dewey's reference to the notion of quality, which will be discussed below, in Sect. 4.3. What is relevant is, rather, the fact that symbols and signs are intrinsically interwoven – remind here the idea of the interlocking of linguistic with non linguistic modes of behavior which Dewey formulates in the article *Peirce's Theory of Linguistic Signs, Thought, and Meaning* – and that they represent two different yet closely connected ways of dealing with things. They are different since, in the case of signs, the actual presence of one of the elements of the relation of significance is needed as the evidential basis for inference, while in the case of symbols, nothing of this kind is required: symbols can be uttered and produced at will. Nonetheless, since the inferred element of the relation of significance is, by definition, not currently present, the response to the observed element varies accordingly to its definiteness as a sign. Without a name for smoke, for instance, the inference from the smoke that is actually present in the agent's visual field to the expected fire would be, as Dewey says, "blind and blundering". This because, symbols are necessary in order to "mark and retain just the qualities of things which are the ground of inference" (LW12, 61). But this implies that, without the capacity to perform a linguistic behavior, the kind of activity that would rely on, and follow from, that "blind" inference would be limited, at best, or exposed to many different risks and dangers.

The last remark provides, I think, the key to correctly frame the issue. The better way to grasp the difference between symbols and signs – as well as the distinguishing interwoveness of meaning and significance in human linguistic life-behaviors – is by paying attention to the different forms of normativity of the activities in which they are employed and involved. According to Dewey, there is no possible normativity outside the semantic dimension of sign – be it natural or artificial. Indeed, in the case of purely mechanical responses to external stimuli there is no standard of evaluation of their rightness and correctness. They are brute, mechanical facts. Normative assessment becomes possible in case of natural signs – or, better said, it becomes possible when behavior is refined enough to take into account the relation that holds between two distinct elements. For reasons that will be evident in a moment, I call this kind of normative assessment 'proto-normativity'. Intuitively, a certain degree of normativity originates when some *present* element or quality is evaluated in light of the consequences that are expected to follow from it. As Dewey writes in *Meaning and Existence* (1928): "events acquire meaning," and meaning here should be understood as Meaning$_B$, "by having their potential consequences identified with them as their properties" (LW3, 85–86).

Though Dewey is not very clear on this point, it is likely that the behavior of animals and lower organisms is subjected to this type of normativity only. Dewey seems to suggest a view of this kind in some passages of his work. In any case, the capacity of human beings to use language dramatically changes the whole scenario. It remains true that human behavior can be assessed in terms of the consequences that are expected to follow from a certain group of qualities of the objects – or, better said, from their perception and manipulation. So, for instance, it is true that "the experienced qualities of sight and smell that constitute a thing 'smoke'", to quote Dewey's own words, are connected to the fire that is their cause, and that the two, taken together, are connected with the burns and suffocation that they may cause (LW1, 61).[17] Nonetheless, precisely because the interwovenness of symbols and signs in linguistic behavior, human beings are never subjected *exclusively* to proto-normativity. As linguistically endowed organisms, human beings are always subjected to a full-fledged normativity.

Such full-fledged normativity consists in the capacity of the agent to assess some aspect of the environment – and, consequently, the course of action that she is going to undertake – in light not only of the restricted set of consequences that are connected to it by the relation of significance, but also of the whole set of possible conceivable consequences that can be derived from the system of relations in which the symbol which refers to that particular quality or event is embedded. Speaking in terms of representative capacity, Dewey remarks, the word smoke "can be related to other meanings in the language-system; not only to that of fire but to such apparently unrelated meanings as friction, changes of temperature, oxygen, molecular constitution, and, by intervening meaning-symbols, to the laws of thermodynamics" (LW12, 58).

The last quotation clearly depicts the difference between the two kinds of normativity as a difference in the scope of the consequences that are connected to the sign. It is worth noting that such difference is not purely quantitative, as it may seem at first glance. The issue at stake is not that of the different scope or extension of the sets of expected consequences; rather, the difference between proto- and full-fledged normativity ultimately comes down to the fact that, contrary to the former, the latter presupposes a conceptual space within which symbols operate and exert their representative or semantic capacity.

Before moving on to address the nature and structure of the conceptual space which makes possible for an artificial sign to have meaning, it is worth noticing that two important consequence follow from Dewey's remarks. First of all, it should be apparent that he does not relapse into the Myth of the Given. According to Dewey,

[17]Recall that the representative capacity of smoke, taken as natural sign of fire, "is highly restricted, for it exists only under limited conditions" (LW12, 58). Dewey's point is that the behavior which relies on the relation of significance has an instrumental character, in a twofold sense: (1) in the sense that the use of a certain element as sign of something else is instrumental to the realization of a certain goal which is causally connected to the first element of the chain, the one which is under our present control; (2) in the sense that the proper function of a sign is that of being used instrumentally, in a concrete course of activity, in light of what is expected to follow from it.

indeed, there is no meaning and normativity outside language, and such equivalence is so strong that the very possibility of the significance of natural signs depends and rests – at least in the case of human organisms – on their capacity to use language. As Dewey explicitly remarks, "it is language, originating as a medium of communication in order to bring about deliberate cooperation and competition in conjoint activities, that has conferred upon existential things their signifying or evidential power" (LW12, 62). It is for this reason that I have chosen the label "proto-normativity" to refer to the normative structure of significance.

Secondly, the thesis of the intrinsic interwoveness of symbols and signs entails that there is *in principle* no semantic difference between the content of symbols and that of signs. As will be outlined in Chap. 4, the aim of inquiry is precisely that of transforming the content of a symbol – or of a set of symbols – into that of the corresponding sign – or set of signs. Now, McDowell has noticed that this conclusion is both a paradox – how is it possible for the mind to catch reality in its net? – and a truism, being evident that with our thoughts and concepts we grasp reality as it is in itself. He refers to Wittgenstein's well known statement that "[w]hen we say, and *mean*, that such-and-such is the case, we – and our meaning – do not stop anywhere short of the fact; but we mean: *this-is-so*" (McDowell 1994, 27; Wittgenstein 2009, §95). And then he formulates that Wittgensteinian insight with the following words: "there is no ontological gap between the sort of thing one can mean, or generally the sort of thing one can think, and the sort of thing that can be the case". This implies, among the other things, that "there is no gap between thought, as such, and the world": "thought can be distanced from the world by being false, but there is no distance from the world implicit in the very idea of thought" (McDowell 1994, 27).

In light of Dewey's approach, in fact, the identity of meaning and significance is neither a paradox nor a truism. Contrary to McDowell and Wittgenstein's, Dewey's metaphilosophical approach is by no means quietist: far from looking for a therapeutic dissolution of philosophical illusions, philosophy should be engaged in the search for explanations, in the strongest continuity with natural and social sciences. Consistently, such an identity of meaning and significance, thought and world, is to be viewed as a natural fact that can be accounted for in naturalistic terms by pointing to the interlocking of linguistic with non linguistic modes of behavior which constitutes the full-fledged normativity of human activity.

2.4 Semantic Externalism and Practical Agreement

In the previous sections I have firstly stressed the centrality of the idea of linguistic behavior for the understanding of Dewey's philosophy of language. Then, I have discussed the metaphilosophical assumptions that lie at the ground of his shift from language to linguistic behavior, understood naturalistically as an activity performed by an organism endowed with a highly refined system of biological structures and processes. Finally, I have pointed out that, properly speaking, there is no meaning outside language. True, there is semantic continuity between language and things – better said, between symbols and natural signs – as a consequence of which the content of a symbol can be projected on, and get somehow incorporated by, the object that is the proper referent of the symbol. In any case, such continuity – which is not problematic from Dewey's perspective since it is accounted for in terms of the activities that are performed by an organism capable of using language – is unidirectional, so to say: it goes from language to things, and not in the opposite direction.

I would like now to focus with greater detail on Dewey's theory of meaning. The starting point of any analysis regarding this issue is the acknowledgment that Dewey's activity-centered conception of language directly entails the idea of the objectivity of meanings. As Dewey remarks, '[m]eanings are objective because they are modes of natural interaction" (LW1, 149). Now, in this context, objectivity means two different yet strongly interrelated things: on the one hand, the fact that meanings are not set and established by a decision of the speaker; on the other hand, the fact that meanings are not private, inner mental states of the speaker. These two conceptions are strongly interrelated in that they both point to the definition of what may be called *practical semantic externalism*, to the effect that the anti-internalist argument is grounded in the structure of activity. The goal of this section is to clarify and articulate this Deweyan thesis.

Let's start with the second issue first, since it can be more easily addressed in light of what has been said thus far. Being modes of life-behavior, it is quite obvious that meanings cannot be confined within the head of the speaker. Dewey is explicit that, insofar as meanings are modes of natural interaction, they also "includ[e] things and energies external to living creatures" (LW1, 149). Recall what Dewey says about experience, namely, the fact that this is a double-barrelled notion which encompasses the thing *experienced* as well as the *experiencing* subject. The same consideration holds in the case of meaning. Dewey's account of meaning is, therefore, strongly externalist. The meaning of a symbol depends on, and is grounded in, the activities that take place in the external world. In their turn, these activities depend on the elements of the environment that are used to achieve the specific goal that those activities purport to reach.

This point can be formulated in a slightly different way, by focusing on the normative element that is entrenched in the notion of meaning. As has been remarked above, the meaning of a symbol coincides with the set of possible consequences that follow from the application of that symbol in all conceivable courses of action. However, the application can be successful or unsuccessful, where the criteria of success do not depend on the internal states of the speaker: as there is no difference, from an internalist perspective, between believing and believing truly, the same holds true in the case of the use and application of symbols. In Dewey's philosophy, success or satisfaction is a highly complex notion: in a provisional manner, it can be said that success in the application of a symbol ultimately depends on whether or not external world responds to the activities of the speaker in a way that satisfactory meets her expectations, as these are ingrained in her habits of behavior. If the speaker perceives smoke and utters, "there is a fire in the room", the meaning of the symbol "fire" is set by the objective or environmental consequences of some possible courses of action that may be undertaken by the speaker. Similarly, the significance of the particular smoke that the agent perceives in the corner of the room is set by the consequences of the courses of action, usually quite stereotyped, that she can undertake in that particular situation. So, for instance, if she went closer to the place from which smoke arises, and could neither find any fire nor sense any acrid smell, then she would be compelled to revise her semantic commitments. As Dewey remarks in a different context, meaning "is independent of the psychical landscape, the sensations and imagery" of the agent and the people concerned in that course of action: meaning is a public affair (LW1, 149).

These last remarks are useful to shed light on an important feature of Dewey's philosophy, which has only been hinted at above. To say that meaning ultimately depends on how external world responds to the activities of the speaker, in a way that satisfactory meets her expectations, does not amount to endorsing some sort of reductionism, which leads to a disguised form of the Myth of the Given. As will be remembered, that was one of the two main criticisms that Black raises against Dewey's philosophy of language. Black charges Dewey with identifying the meaning of a symbol with its nonverbal counterpart, which exists "independently of the symbol whose meaning it is alleged to be" (Black 1962, 519; see above). From what has been said, it should be clear why Black's criticism is misplaced. Indeed, Dewey does not think that a nonverbal counterpart of meaning exists, to which the latter can be traced back; rather the contrary, what counts as a satisfactory response to a certain course of action depends on the standards and criteria internal to that very activity, which is linguistic through and through, and consists of the set of possible activities that an organism complex enough to master the use of language can undertake.

With this in mind, we can now move on to discuss the other pillar of Dewey's practical semantic externalism, namely, the idea that objectivity of meaning entails the publicity of language. In the wake of Wittgenstein's private argument, it is usually admitted that language is constitutively public, on the basis that a private language would lack the normative constraints necessary for linguistic signs to have meaning. Dewey too endorses the idea of the impossibility of a private language, and stresses the fact that language is a social phenomenon. He also agrees with Wittgenstein that the issue at stake is the idea of the sociality of normativity. However, Dewey's private argument, so to say, is significantly different from Wittgenstein's one, even though they share the same inspirations and, in the last account, reach the same conclusion.

In Dewey's philosophy, that conclusion – which amounts to arguing that the meaning of a symbol cannot be established by any act or set of acts of decision – follows directly from acknowledging that the criteria that are needed to evaluate the success of the application of a symbol are objective in that they are related to the success of the course of activity in which that symbol is employed. "Being a name or *linguistic* sign," Dewey remarks, "is a matter of the performance of behavioral operations" (LW16, 307–308). To be able to master the use of language means to have the capacity to take something – primarily, a sound – as a sign of something else, which is *conventionally* related to it. It is not the case that first something is established as a sign, and then it exerts the function of referring to a specified subject-matter. Rather the contrary, these are two aspects of one and the same transaction. Dewey formulates this point by saying that "[o]ne and the same set of behavioral operations determines, in strict conjunction or conjugation, *sign* and *referent*" (LW16, 308). Or, in more technical terms, "the behavioral operations to which the name is given enact at one and the same time and place the referr*ing* that in one aspect of the total function constitutes being a sign and in another aspect of the same function an 'object' as referent" (LW16, 308). Here the functional identity of sign and object is explicitly stated, and the semantic identity thesis (SIT) is articulated with respect to language, thus providing a solid ground to Dewey's semantic analysis.

Now, the behavioral operations that constitute the function of referring cannot be performed by a single, private agent. This for two different reasons, that Dewey is careful to keep separate. The first reason has to do with the genesis of language; the second with its proper functioning.

I won't deal extensively with the first argument, since a systematic reconstruction of Dewey's theory of the origin of language is a difficult task – mainly because of its scattered and sketchy nature – which falls outside the scope of the present work. It suffices to note that in the fifth chapter of *Experience and Nature* Dewey outlines a social account of the origin of language which centers around the idea of communication. Recall that, in Dewey's perspective, the problem of the origin of language amounts to clarifying the genesis of the behavioral operations that make possible for an organism to use language in its courses of activity.

Now, Dewey holds that meaning originates in the context of communication, when two persons participate in a shared activity. A shared activity is constitutively different from a private course of action in that, in the former case, the agent is asked to take the stance of the other participant – as Dewey remarks, communication means "making of something common" (LW12, 52) – in order to understand how to contribute to the joint action: she has therefore to adopt a more complex perspective, as a consequence of which she can perform the action that the other participant expects her to do. From Dewey's point of view, this is the fundamental – 'fundamental' here should be understood genetically rather than functionally – source of the normativity of language: it stems from the reciprocal expectations of the agents engaged in a shared activity. It is only in a second moment that meanings thus originated are projected on things.[18]

[18]Dewey is explicit on this point: "[p]rimarily meaning is intent and intent is not personal in a private and exclusive sense. A proposes the consummatory possession of the flower through the medium or means of B's action; B proposes to cooperate or act adversely in the fulfillment of A's proposal. Secondarily, meaning is the acquisition of significance by things in their status in making possible and fulfilling shared cooperation. In the first place, it is the *motion and sounds* of A which have meaning, or are signs. Similarly the movements of B, while they immediate to him, are signs to A of B's cooperation or refusal. But secondarily the thing pointed out by A to B gains meaning.

The other argument in support of the thesis of the constitutive publicity of language is more interesting for my purposes, since it brings to the fore the relations that Dewey establishes between semantic externalism and practice. This argument is presented in the third chapter of the *Logic*, as a corollary of the discussion of a passage from Ogden and Richards's *Meaning of Meaning* – who, in their turn, drew it from the book *Among Congo Cannibals*, written by J. H. Weeks.

Dewey was impressed by the following incident, which he takes to be evidence of the fact that symbols function only in a shared context of activity. Better said, to use Dewey's own words, it should be taken as evidence that "words mean what they mean in connection with conjoint activities that effect a common, or mutually participated in, consequence" (LW1, 59). The passage is quite well-known, but it is useful to quote it in its completeness:

> A visitor in a savage tribe wanted on one occasion "the word for Table. There were five or six boys standing around, and tapping the table with my forefinger I asked 'What is this?' One boy said it was *dodela*, another that it was an *etanda*, a third stated that it was *bokali*, a fourth that it was *elamba*, and the fifth said it was *meza*." After congratulating himself on the richness of the vocabulary of the language the visitor found later "that one boy had thought he wanted the word for tapping; another understood we were seeking the word for the material of which the table was made; another had the idea that we required the word for hardness; another thought we wished the name for that which covered the table; and the last . . . gave us the word *meza*, table." (LW12, 59).

The issue at stake here is the proper understanding of the notion of convention or agreement. Quite traditionally, Dewey distinguishes between signs and symbols on the basis of the difference between the naturality of the former and the artificiality of the latter. I have stressed above that to a closer look Dewey's views are actually much more complicated than that, but there is no doubt that, to a certain extent, the traditional account has some degree of plausibility and reliability. Indeed, it is clear that the right translation of the word 'table' as 'meza' ultimately depends on the fact that in the language of the savage tribe 'meza' is used to refer to tables. It is a matter of convention since symbols do not have their meanings as a natural property.

The passage quoted above aims precisely to highlight this feature of language. With that example Dewey purports to show that the meaning of a symbol is completely indeterminate outside the context of its application. No matter how much empirical evidence is or can be provided, the meaning of a linguistic sign cannot be established independently of the context of the shared activity in which it is used. So, in the case under discussion, the mere act of tapping the table is not enough to fix the meaning of the word; something more must be added. Indeed, each participant to the activity has the same amount of data, and yet no agreement is reached. Clearly, this 'something more' is not a surplus of facts; it must be the understanding of the specific activity in which the speaker is engaged, and in which the hearer is asked to participate. As Dewey explicitly remarks, "[t]he act of tapping in the illustration

---

It ceases to be just what it brutely is at the moment, and is responded to in its potentiality, as a means to remoter consequences" (LW1, 142).

was isolated from any such situation;" it was "no part of *communication*, by which alone acts get significance and accompanying words acquire meaning" (LW12, 59).

I will deal extensively with the consequences of this thesis for Dewey's theory of inquiry in the following chapter, where I will address the theory of normativity that lies at the basis of the notion of situation. As for now, let's go back to the issue of the nature of convention or agreement. From what has been said thus far, it is safe to say that, according to Dewey, meaning rests on agreement, provided that agreement – the kind of agreement that makes it possible for an organism to fix the meaning of the symbols used in communication – be understood as practical agreement, namely, agreement in the shared activity in which two or more persons participate. Dewey formulates the insight in the following way:

> The physical sound or mark gets its meaning in and by conjoint community of functional use, not by any explicit convening in a 'convention' or by passing resolutions that a certain sound or mark shall have a specified meaning. Even when the meaning of certain legal words is determined by a court, it is not the agreement of the judges which is finally decisive. For such assent does not finish the matter. It occurs for the sake of determining future agreements in associated *behavior*, and it is this subsequent behavior which finally settles the actual meaning of the words in question. Agreement in the proposition arrived at is significant only through this function in promoting agreement in action (LW12, 52–53).

Dewey's idea is that while 'linguistic' convention is unsuccessful in determining the meaning of linguistic signs, practical agreement can succeed in achieving that goal since it does not consist in the mere decision to use the words in a certain ways, but rather amounts to a reciprocal agreement in the activities (linguistic and non-linguistic) that are commonly undertaken by their participants. As Dewey puts it, "[i]t is an error to suppose that the misunderstanding is about the meaning of the *word* in isolation, just as it is fallacious to suppose that because two persons accept the same dictionary meaning of a word they have therefore come to agreement and understanding" (LW12, 53).

The two errors share a common core: in both cases, the complexity of linguistic activity is somehow underestimated. On the one hand, those who believe that the meaning of a linguistic sign can be determined in isolation from its context of application completely overlook the structural difference between natural and artificial signs. In so doing, they are led to conclude that there is a one-to-one correspondence of words and things, as a consequence of which the semantic properties of language are 'metaphysicized' and made naturalistically unaccountable.

On the other hand, those who argue that purely linguistic convention suffices to determine the meaning of linguistic signs are committed to an inconsistent view. According to this reading, indeed, agreement is to be reached once for all at the very beginning of the process of communication. This would entail, once again, that meaning has to be determined independently of linguistic life-behavior. And it would also entail that language should be grounded in something which is pre-linguistic, which only secures its semantic properties. Consequently, those assumptions, taken together, either lead to reinstate the Myth of the Given or are explanatory idle since they leave the possibility of agreement – the fact that the 'contracting parties' understand the kind of activity in which they commonly participate – wholly unaccounted.

Dewey's solution to this problem is to hold steady that, to quote his own words, "agreement and disagreement are determined by the consequences of conjoint activities". His approach is made of two main steps. First of all, Dewey shifts attention from the *fact* of linguistic agreement and disagreement – agreement (or disagreement) on the meaning of a word – to their *possibility*. Indeed, Dewey is less interested in the actual agreement or disagreement than in the conditions that make them possible, in the conviction that what has to be explained is the structure of the logical space within which single, specific linguistic activities are possible.[19]

Secondly, agreement and disagreement are not individual and autonomous facts, which can be analyzed in themselves, independently of the totality of *linguistic* behavior, which provides the context and the criteria in whose light only they can take place. If it were so, a case of genuine agreement could not be told apart from those cases in which agreement happens by chance. Agreement in action is reached when the consequences that are produced by using a certain word or a certain language succeed in satisfying the expectations of the agents involved in the shared activity, *and* that satisfaction would not be possible aside from those acts.

The meaning of a symbol is, therefore, not only contextual; it is also holistic. As will be seen in a moment, Dewey's practical holism is rather idiosyncratic, and presents many features of originality. For now, it is sufficient to notice that, taken together, contextualism and holism ground Dewey's global semantic anti-representationalism.[20] The shape of that conception will be investigated in the following section.

2.5 The Pragmatist Account of Concepts

After clarifying the main features of Dewey's philosophy of language, it is time to turn attention to his fine-grained analysis of concepts. First of all, a terminological remark. Concept is one of the terms that Dewey employs to refer to the content of linguistic signs, alongside with meaning, conception, and idea – the latter being used

[19]Here is what Dewey writes in this regard: "Reference to concord of consequences as the determinant of the meaning of any sound used as a medium of communication shows that there is no such thing as a *mere* word or *mere* symbol. The physical existence that is the vehicle of meaning may as a particular be called mere; the recitation of a number of such sounds or the stringing together of such marks may be called mere language. But in fact there is no word in the first case and no language in the second. The activities that occur and the consequences that result which are not determined by meaning, are, by description, only physical. A sound or mark of any physical existence is a part of language only in virtue of its *operational* force; that is, as it functions as a means of evoking different activities performed by different persons so as to produce consequences that are shared by all the participants in the conjoint undertaking" (LW12, 53).

[20]On this point, see Gronda and Turbanti (2017).

when he wants to emphasize the logical import of concepts. So, while life-behavior is a wholesale notion, which encompasses the totality of the environment plus the organism, concept refers to, and singles out, a specific aspect of the situation. In other words, while life-behavior is a zero word, concept is a proper term.

Dewey's allegiance to pragmatism has often been matter for discussion. Now, I believe that if one is searching for a clear statement of Dewey's commitment to that tradition of thought, his theory of concepts is the right place to look to. Traditionally, pragmatism has been very much concerned with the nature of concepts. Recall Peirce's pragmatic maxim in its original formulation: "[c]onsider what effects, that might conceivably have practical bearings, we conceive the object of our conception to have. Then, our conception of these effects is the whole of our conception of the object" (Peirce 1986, 266). The goal of pragmatic maxim is acknowledged by Peirce to be that of providing a third degree of clarity of the concepts employed in an experimental context. A concept is here identified with its meaning, and its meaning consists of all the practical effects that are brought about by any possible activity that concerns the object of the concept.

Dewey undoubtedly accepts this line of thought. To my knowledge, his most explicit adhesion to the pragmatist approach to the nature of concepts can be found in the fifth chapter of *The Quest for Certainty*. The goal of that chapter, whose title is *Ideas at Work*, is to present an account of concepts that fits within, and corroborates, Dewey's practical turn, as is outlined in the last chapter of the book, entitled *The Copernican Revolution*. What is interesting to note is that, in this text, Dewey couches his *pragmatist* theory of concepts in the language of the operationalist program as formulated by Bridgman and Eddington.

Dewey's reference to Bridgman and Eddington may look surprising. Actually, the space that Dewey devotes to analysing and discussing their views has no equal in his work. In addition, as clearly emerges from his correspondence with Sidney Hook, the reasons why in *The Quest of Certainty* – recall that that book grew out from Dewey's Gifford Lectures, delivered in 1929 – Dewey deals so extensively with operationalism are to some extent contingent (1929.01.04, J. Dewey to S. Hook). Nonetheless, it is hardly possible to deny that there are objective reasons that motivate Dewey's interest in the operationalist philosophy of science. As is well known, Bridgman's operationalist account of scientific concepts connects, in an intrinsic way, the meaning of scientific concepts with a specific set of experimental operations. Clearly, this approach is in strong agreement with Dewey's practically-oriented philosophy of language. No surprise, therefore, that Dewey approvingly quotes the following passage drawn from Bridgman's *The Logic of Modern Physics*:

> To find the length of an object, we have to perform certain physical operations. The concept of length is therefore fixed when the operations by which length is measured are fixes; that is, the concept of length involves as much as and nothing more than the set of operations by which length is determined. In general, we mean by any concept nothing more than a set of operations; the concept is synonymous with the corresponding set of operations (LW4, 89; Bridgman 1927, 5).

The same idea Dewey finds expressed in Eddington's *The Nature of the Physical World*. In particular, he supportively reports the following statement: "[i]t is now recognized that these [words such as length, angle, velocity, force, and so on] should be defined according to the way in which we recognize them when actually confronted with them, and not according to the metaphysical significance which we may have anticipated for them" (LW4, 89; Eddington 1928, 254–255). Dewey's point, following Eddington, is that the meaning of scientific concepts is to be defined and understood in terms and in light of the experimental method by which they are established, rather than in terms and in light of some metaphysical unwarranted assumptions. This means that "concepts are recognized by means of the experimental operations by which are determined"; and, even more clearly, that "operations define and test the validity of the meaning by which we state natural happenings" (LW4, 90).

Dewey's point is that these operationalist insights should be read in strong continuity with Peirce's theory of meaning, as formulated in *How to Make Our Ideas Clear* – namely, the view that "the sole meaning of the idea of an object consists of the consequences which result when the object is acted upon in a particular way" (LW4, 90). In Dewey's view, what these two accounts of meaning, the operationalist and the pragmatist, have in common is the idea that the meaning of a scientific concept – but this holds true for *any* concept whatsoever, as will be made clear in the next section – is to be cashed out operationally. Contrary to any metaphysical approach, the meaning of a concept is to be traced back to the concrete differences it makes in practice, the background assumption being that no conceptual distinction is so complex and refined that cannot be revealed and expressed in some course of activity.[21] So, pragmatism and operationalism converge in an antirepresentationalist theory of concepts, to the effect that a concept does not consist in a representation of some portions of the environment – whose criteria of validity lie in the faithfulness and adequacy of the representation – but is rather a rule for promoting certain controlled transformations of the environment so as to bring about certain expected consequences. Meaning is therefore the resultant of a complex and multifaceted networks of operations by which and through which the organism copes and interacts with the environment.

[21] Actually, there are reasons to believe that Dewey's position is more complex than I have so far indicated. In *Experience and Nature* he sometimes distinguishes between cognitive and esthetic or literary meanings – see, for instance, (LW1, 245). In some other cases, he explicitly restricts his operational analysis to intellectual meanings only. See, for instance, the following passage: "[i]n responding to things not in their immediate qualities but for the sake of ulterior results, immediate qualities are dimmed, while those features which are signs, indices of something else, are distinguished. A thing is more significantly what it makes possible than what it immediately is. The very conception of cognitive meaning, intellectual significance, is that things in their immediacy are subordinated to what they portend and give evidence of. An intellectual sign denotes that a thing is not taken immediately but is referred to something that may come in consequence of it. Intellectual meanings may themselves be appropriated, enjoyed and appreciated; but the character of intellectual meaning is instrumental" (LW1, 105). These remarks seem, therefore, to pave the way for a more careful and nuanced reading of Dewey's theory of concepts, according to which what can be formulated in operational terms does not exhaust the *entire* meaning of the concept, but only its cognitive or intellectual import.

As often with Dewey, however, things are much more complicated than it may seem at first glance. What is at stake here is the degree of identity, so to speak, that Dewey claims to hold between pragmatism and operationalism. In the context of the passages quoted above, indeed, Dewey stresses the continuity between pragmatism and operationalism to such an extent that he goes as far as stating that their "*logical* import is identical" (LW4, 90). He also states that, precisely because of the identity in logical import between the two approaches, it is possible to formulate the pragmatist theory of concepts in terms of what he calls "operational thinking", thus getting rid of the "ambiguities in the notion of pragmatism" (LW4, 90). Nonetheless, I think it is rather easy to see that such an identification – no matter how explicitly claimed – does not correspond to Dewey's actual position. There are at least two relevant differences between the two approaches, of which Dewey is certainly aware.[22]

First of all, pragmatists and operationalists have two slightly different conceptions of what counts as operational thinking. Indeed, while Bridgman and Eddington equate meaning with a particular set of *operations*, pragmatists identify it with the *relations* that hold between a specific set of operations performed by the agent and the objective, environmental consequences that are expected to follow from them. Recall the words with which Dewey formulates Peirce's theory of meaning: "the sole meaning of the idea of an object consists of the *consequences* which result when the object is acted upon in a particular way" (LW4, 90; italics mine).

The shift from operations to the consequences of those operations is philosophically relevant. On the one hand, it dispels any risk of subjectivism. Though only surreptitiously, speaking of operations may convey the idea that we are concerned exclusively with the activities that the agent performs – or that she can perform, if we are interested in preserving the modal character of the concept. This is clearly not what Dewey has in mind since operations – in this specific sense – are abstractions from a broader reality, which encompasses the subject as well as the object, the organism as well as its environment. So, by insisting on the notion of consequences, Dewey stresses once again the primacy of activity over its components.

On the other hand, the insistence on the notion of consequence confirms and supports Dewey's general theory of signs in a way the notion of operation cannot do. Indeed, it is precisely because the meaning of a concept consists of, and is determined by, the consequences of a group of operations performed by the agent in a particular context that meanings can be objective in the sense required by Dewey's philosophy of language. While operations do not bear in themselves any explicit relation to their expected results, consequences of actions are, by definition, publicly ascertainable, and, accordingly, provide the ground for agreement in activity.

[22]A third difference may be added, which concerns the scope of the operationalist account of concepts. While Eddington and Bridgman are concerned exclusively with the meaning of *scientific* concepts, pragmatists – and Dewey in particular – believe that the operationalist account can be extended to hold for any type of concepts.

In any case, it is important not to exaggerate the differences between pragmatism and operationalism on this issue. At the end of the day, one might argue that speaking of operations is, at worst, a simplified and inaccurate way of formulating the insight that Dewey expresses in the language of consequences. I do not agree with this line of reasoning, but I concede that it has some plausibility. The problem with that reading is that it ends up trivializing a philosophically interesting point. The operationalist account can be easily accommodated within Dewey's framework, provided that the former is acknowledged as dealing with a relevant component of meaning, which is got by abstraction from the complex totality. More explicitly, I think that the fundamental tenets of operationalism are preserved in Dewey's philosophy of science by being translated into the technical notion of universal propositions. I will deal with this aspect of Dewey's theory of inquiry later in Chap. 4.

On the contrary, the second main difference between Deweyan pragmatism and operationalism – which concerns the structure of *scientific* concepts – is sign of a radical, and more fundamental, disagreement on the nature of meaning. As has been remarked above, according to Bridgman, scientific concepts are entirely determined by a set of operations. Bridgman is explicit on this point: "*the concept is synonymous with the corresponding set of operations*" (Bridgman 1927, 5). It is clear that the relation that Bridgman sees between concepts and sets of operations is semantic, as is revealed by his use of the notion of synonymy. So far so good, one might say, provided that the operations are linguistic through and through. Otherwise, it would imply a relapse into the Myth of the Given.

One of the consequences that Bridgman draws from that assumption is that the meaning of a scientific concept is *entirely* determined by a specific and definite set of operations. This implies that any change in the set of operations also implies a change in the concept. Bridgman is explicit on this point, as is clear from his discussion of the concept of length:

> *In principle* the operations by which length is measured should be *uniquely* specified. If we have more than one set of operations, we have more than one concept, and strictly there should be a separate name to correspond to each different set of operations [. . .]. If we want to be able to measure the length of bodies moving with higher velocities such as we find existing in nature (stars or cathode particles), we must adopt another definition and other operations for measuring length, which also reduce to the operations already adopted in the static case. This is precisely what Einstein did. Since Einstein's operations were different from our operations above, *his "length" does not mean the same as our "length"* (Bridgman 1927, 10–12).

The disagreement between pragmatism and operationalism lies here. Dewey is strongly critical of any form of meaning fixism, to the effect that a concept is synonymous with a fixed and unchanging set of operations. Beware that the reason why Dewey is dissatisfied with the operationalist approach does not abide in the latter's identification of meaning with a set of operations. Actually, Dewey endorses a particular version of this view. Rather, the main reason of his dissatisfaction lies in the identification made by operationalists of meaning with a *fixed* set of operations. Indeed, meaning fixism supports a strongly instrumentalist view of scientific concepts – and scientific objects – that Dewey is not willing to endorse.[23] Contrary to Bridgman, Dewey is committed to the thesis that scientific concepts have a theoretical autonomy that is at the basis of their capacity to create new knowledge. Such a theoretical productivity seems to be wholly unaccountable within an operationalist framework.

With all the due differences, Dewey's point is similar to the one made by Hempel with his theoretician's dilemma. The problem with which Hempel is concerned is that of understanding whether "the theoretical detour, through a domain of not directly observable things, events, or characteristics cannot be entirely avoided" (Hempel 1965, 185–186). Paraphrasing Hempel, it can be said that Dewey's problem is that of understanding whether a theoretical detour through a domain of not directly fixed and pre-established consequences of some set of operations is possible or not. According to Bridgman, such detour is not possible since the meaning of a concept is fixed once for all, at the very beginning of inquiry. According to Dewey, on the contrary, the fact that through the process of inquiry scientific concepts evolve brings indisputable evidence for its possibility.

Dewey is not very clear about how the process of conceptual evolution may happen. In *The Quest of Certainty*, he states that "[t]his conception [the concept of length] in connection with other operations, such as those which define mass and time, become instruments by means of which a multitude of relations between bodies can be established" (LW4, 100). Here it seems evident that Dewey has in mind the possibility that different operations may interact one with the others, thus generating new and more complex modes of dealing with objects. In the *Logic* Dewey expresses the same point in a slightly different way. After emphasizing the structural differences between signs and symbols, and stressing the fact that by being embodied in an artificial sign meaning "is *liberated* with respect to its representative function", he writes:

> [Meaning] is no longer tied down. It can be related to other meanings in the language-system; not only to that of fire but to such apparently unrelated meanings as friction, changes of temperature, oxygen, molecular constitution, and by intervening meaning-symbols, to the laws of thermodynamics (LW12, 58).

Leaving aside the differences between the two formulations, it is clear that they both support a holistic account of the meaning of scientific concepts. In both cases, what Dewey wants to stress is that the meaning of scientific concepts is open to a continuous refinement, as a consequence of their being employed in more and more complex and articulated courses of action. So, for instance, the concept of length – which is originally defined by some set of activities – changes and evolves when it is used to measure the length of bodies moving with higher velocities. By entering in connection with other meanings and operations, the concept of length – or the

[23]This statement may look surprising since Dewey's philosophy of science is usually taken to be a paradigmatic form of instrumentalism. As I will argue in Chap. 5, such an interpretation is mistaken.

concept of smoke, to which Dewey refers in his *Logic* – is progressively refined, and acquires a more and more precise and definite meaning.[24]

But what kind of holism is Dewey committed to? In light of Dewey's commitment to pragmatism, one would therefore be tempted to conclude that Dewey's holism should be understood as a confirmation holism, which ultimately depends on a verificationist theory of meaning which has been traditionally associated with that tradition of thought.[25] This reading has clearly some plausibility. It is true, for instance, that according to Dewey meaning is related to its method of verification. It is also true that Dewey often uses a verificationist language, and that, within this framework, reference to observable entities is a necessary condition for a concept to be meaningful. Now, since there is no one-to-one correspondence of words and things, scientific concepts have observational implications – which ultimately provide the criteria for their verification – only if joined with other concepts of the theory.

It is worth noting, however, that, all this notwithstanding, Dewey's holism cannot be boiled down to confirmation holism. Indeed, one of the distinctive traits of Dewey's holistic account of concepts is the thesis that scientific concepts acquire more and more meaning as a consequence of their being included in more and more complex scientific theories. While confirmation holism goes top-down, so to say, the idea of "liberation" of meanings goes in the other direction, bottom-up. The productive character of idealization – the fact that, by means of idealization, a common-sense concept like fire is "related to other meanings in the language system" such as "friction, changes of temperature, oxygen, molecular constitution, and by intervening meaning-symbols, to the laws of thermodynamics" – cannot be accounted for in terms of confirmation holism (LW12, 58).

I think that the best way to characterize Dewey's holism is to read it as a particular form of inferential holism. As an use-centered approach, inferential holism is committed to the view that there is an essential and constitutive relation between meaning and some aspects of the agent's use of the corresponding concept. Traditionally – think for instance at Brandom's inferentialism – the use which is relevant to the determination of the meaning of a concept is identified with its inferential role. As Brandom points out, the key of an inferential approach to semantics is that of conceiving the content of any non-logical concept as "exhaustively constituted

[24]A remark is in order here. As any holistic approach, Dewey's theory of concepts blurs the clear-cut distinction between semantics and epistemology. However, Dewey's version is not, strictly speaking, epistemic. It seems clear from what has been said that he does not endorse the view that the meaning of a concept consists in the webs of beliefs had by the agent when using that concept – or the corresponding word. As will be shown in the next chapter, Dewey strongly rejects the language of belief as he considers it ingrained in a now completely untenable subjectivist epistemology. In any case, the point that Dewey wants to make has nothing to do with what an agent happens to believe; rather, it concerns the totality of the objective consequences of her experimental activities.

[25]On the relationship between pragmatism and verificationism, see Misak (1995), chapter 3. See also the interesting (Putnam 2017).

by the material, non logical *inferential* connection between those circumstances and consequences," i.e., between the circumstances and consequences of application of those concepts (Brandom 2007, 654). So, for instance, take the concept of temperature:

> The content of a concept such as temperature is, on this view, captured by the constellation of inferential commitments one undertakes in applying it: commitment, namely, to the propriety of all the inferences from any of its circumstances of appropriate application (Brandom 2007, 654).

Dewey is more radical on this point. Instead of focusing on the *inferential* role of concepts, he rather focuses on what may be called their *experimental* role. He thus identifies the meaning of a concept with the totality of the *possible* roles that a concept can play in the activities that an agent can undertake. And Dewey is explicit that possible here is to be taken as referring to existential activities. Indeed, he openly acknowledges that there is an "ambiguity in the term 'possible' operation": possible may be interpreted in a mathematical or in a physical way. Now, he stresses, "[i]ts primary meaning is actually, existentially, possible" (LW4, 127). Possible, therefore, stands in a middle position between actually performed operations and those that are merely possible, independently of any reference to existence. "Any idea as such," Dewey writes "designates an operation that *may* be performed, not something in actual existence" (LW4, 127–128). Accordingly, the operation which defines the meaning of the concept does not need to be an actual operation, but it has nonetheless to be a concrete, experimental – and not purely symbolic – operation.

I will come back to this issue – the comparison between Dewey and Brandom's inferentialist accounts – in the next chapter. For now, I want to stress that such shift from inferential role to what I have called experimental role has at least three remarkable philosophical consequences. First of all, since the process of idealization makes it possible for an agent to use the concepts in an increasing number of activities, their meaning is not fixed once for all, but is open to constant redefinition. Dewey's holism is, therefore, strongly practice-oriented: the content or meaning of any concept is dependent on the practices in which the concept is applied. As will be remembered, this is precisely the point in which Deweyan pragmatism and operationalism part their ways. Consequently, this reconstruction of holism is – at least to this extent – in agreement with Dewey's desiderata.

Secondly, Dewey's holism has a strong 'articulative' component that gives it a quite idiosyncratic character. With 'articulative' I mean to refer to a specific implication of Dewey's idea that meanings get liberated in their representative function by being embodied in artificial signs. As hinted at above, the liberation of meanings – in their representative function – is a process of semantic refinement or idealization that issues from their employment in different, more and more complex activities and life-behaviors. It is this process of refinement that I label *articulation*.

This notion of articulation – and, consequently, the notion of meaning-liberation on which it is grounded – can be understood either in a strong or a weak sense. In the strong sense, what is articulated is 'already there', from the very beginning, in the content that undergoes the process of articulation. According to this reading, articulating amounts to making explicit what is only implicitly formulated at the early stages of the process.[26] In the weak sense, articulation is to be understood in a more relaxed manner. From this viewpoint, the process of articulation does not merely consist in making the implicit explicit; rather, it is taken as capable of creating new meanings which accrue on the already existing ones.

Even though there is some evidence in support of the strong expressivist reading, especially in Dewey's early works, I take Dewey to be committed to the weaker reading.[27] Several arguments can be advanced in favor of my interpretation. I will list two of them. The first one runs as follows. Take the strong expressivist reading as valid. As said, this means that the goal of the process of articulation is to make the implicit explicit. So, it seems to follow from that assumption that – to use Dewey's example – the notion of oxygen and the content expressed by the laws of thermodynamics are implicit in the concept of fire. This conclusion has some plausibility since it is difficult to deny that there is some kind of objective connection between fire, on the one hand, and oxygen and the laws of thermodynamics that explain its behavior, on the other hand. I concede this point, and I think that a sound conception of articulation must acknowledge it. However, what kind of *semantic* relation is this? More precisely, in what sense is it possible to say that the laws of termodynamics are implicit – not in a broad and relaxed sense, but rather in a strong and technical one – in the *meaning* of the concept of fire?

I think nobody is willing to deny that the concept of fire can be used quite independently of the mastery of the concepts of oxygen, combustion, and so on; not least because the latter, being scientific concepts, may turn out to be false, while it is difficult to argue that the same fate can be shared by the common-sense concept of fire. Nonetheless, under the strong expressivist interpretation, it seems that any meaning that the concept may acquire has to be recognized as the expression of something that was implicit in the original meaning – quite irrespectively of its being valid or not. So, for instance, the concepts of phlogiston and oxygen are to

[26]An even stronger reading might be advanced, to the effect that the articulated meaning provides the necessary conditions for the mastery of the implicit meaning. I will not discuss this position because it is clearly untenable from a Deweyan perspective.

[27]For instance, in his *Psychology* Dewey accepts the strong expressivist reading. See, for instance, the following two passages, drawn from Chapter 8 of that book. In the context of his analysis of the function of thinking, Dewey firstly writes: "[j]udgment [. . .] stands in a twofold relation to conception. In one aspect, its analytic, it is based on the concept and develops it; in the other, its synthetic, it returns into the concept and enriches it, by connecting some new element with it. Reasoning [. . .] stands in a like relation to judgment, and therefore to conception. It is based on judgment, for it takes two or more judgments, that is, affirmations of relations, and analyzes them to discover the common or identical relation which unites them. And it expresses this in the form of a new judgment" (EW2, 199–200). And then he concludes: "[p]hilosophy [. . .] is, therefore, no new kind of knowledge, but is the conscious development of what is unconsciously at the hear of all knowledge – the presence of unity in variety" (EW2, 202). I do not want to enter in detail here into the problem of understanding Dewey's early philosophy. Suffice it to note that Dewey quite clearly endorsed what I call the strong conception of articulation in his early texts, which he then decided to abandon in his later works.

be viewed as both implicit in the concept of fire, since the only difference between them seems to consist in the validity of the scientific theory to which they belong. But since any scientific theory is, at best, only approximatively true, it seems fair to say that there is no solid semantic ground to conclude that the meaning of the concept of oxygen is truly an expression of the meaning of the concept of fire, while the meaning of the concept of phlogiston is not.[28]

The second argument against the strong expressivist account of articulation is 'metaphysical' – in Dewey's sense of the word – and revolves around the autonomy and reciprocal independence of the life-behaviors that human beings can undertake. Accordingly, this argument is more clearly connected with the practice-oriented character of Dewey's holism. Here is the argument. From a pragmatist perspective, the meaning of a concept is defined by the consequences of all the possible operations in which that concept is applied. Now, life-behaviors are independent one from the others. Take, once again, the example given by Dewey: I can use smoke to infer the presence of fire, as well as to investigate the properties of combustion or to assess the validity of the laws of thermodynamics. Those three life-behaviors are independent in that they have different conditions of success and satisfaction. Consequently, the strong interpretation of the implicit-explicit relation does not grasp the semantic relation that hold between the different aspects of the meaning of a concept.

The last remarks lead directly to the third point that I would like to make. Dewey's practice-oriented holism has the conceptual resources to avoid some unfortunate consequences of traditional holism. One of the main criticisms addressed to holism is that, if it were true, any change in knowledge would imply *ipso facto* a change in meaning. Clearly, this is a very strong and demanding assumption which flies in the face of our common-sensical intuitions about meaning. Now, even though Dewey is a holist of some sort, I think he is not compelled to accept that thesis. The point is that the reciprocal independence of the various life-behaviors which the agent can undertake supports a distinction between those core-properties that constitute the meaning of the concept and some other properties that, though affecting the *totality* of its meaning, are not so central and fundamental. As will be highlighted in the next section, this distinction is intrinsically interwoven with Dewey's characterization of the relationship between science and common sense.

[28]A remark is needed here in order to avoid confusion. The point at stake is not simply that the strong expressivist account of articulation is explanatory idle; the problem is, rather, that it purports to introduce a strong semantic relation between meanings which should make some difference as to our accessibility to them; in so doing, it ends up arguing that it should be possible to solve epistemic problems through and by semantic considerations. Even within a holist framework, such a view is so problematic as to be untenable.

2.6 Common Sense and Scientific Concepts

In the previous section I have laid out two arguments in support of what I have called the weak conception of articulation. Those two arguments were mainly negative: their aim was to show that the strong conception is untenable, and to highlight why it fails to provide a satisfactory account of the process of meaning-liberation. It is now time to turn attention directly to the explanatory advantages of the weak conception.

Before starting the discussion proper, it is useful to recapitulate briefly the facts of the matter. The theoretical issue related to the notion of articulation has to do with the kind and scope of the relations that, within a holistic framework of analysis, constitute the meaning of a concept. Notoriously, the problem with semantic holism is twofold: on the one hand, by blurring the distinction between semantics and epistemology, it seems to lead to the strongly counterintuitive conclusion that any change in belief implies a change in meaning; on the other hand, it seems to entail a sort of semantic instability, which follows from the assumption that anytime a new piece of information is added to the stock of existing knowledge, the meaning of the corresponding concept changes.

Any defender of semantic holism is, therefore, compelled to find a way to reject those conclusions without giving up the idea that the meaning of a concept consists in the net of relations that it entertains with other elements of the logical space. In other words, the optimal stance for a semantic holist is that which enables her to say that the meaning of a concept is modified – though imperceptibly, at least most of the time – anytime new information is added, *and* that there is a sort of kernel of meaning which remains relatively invariable through the various changes, thus providing the identity conditions of the concept. I believe that Dewey's distinction between science and common sense – or, better said, between those life-behaviors which are guided by the search of knowledge for the sake of knowledge and those which are concerned with the practical affairs of everyday life – is intended to satisfy these two demands.

Dewey's argument runs as follows. As has been said, from a pragmatist perspective the meaning of the concept of an object consists entirely of the consequences which result when the object is acted upon in a particular way (LW4, 90). A concept is always employed in a mode of activity, where it works by pre-establishing the kinds of response that are expected from the environment as a consequence of the actions performed by the agent. So, for instance, the concept of water states that if the corresponding object is drunk whenever one is thirsty, it will quench the thirst. Quenching the thirst is thus part of the meaning of the concept of water: the relation between the act of drinking and quenching the thirst has acquired a semantic import.

According to this perspective, the more the relations between the actions performed by the agent and the consequences of the environment, the more defined is the meaning of the corresponding concept. So, when it is observed that, put in water, salt dissolves, the concept water enters into a new and different web of relations which changes its meaning. From that moment on, the concept of water means the liquid in which salt dissolves, and not only the liquid that quenches the thirst.[29]

Clearly, when the concept is employed in a refined context of activity, such as the one carried out in a scientific laboratory, the consequences that follow from its adoption as a rule of action are more complex, and less easily predictable, than those that are brought about in practical life-behaviors. As Dewey remarks, the distinguishing character of scientific operations is that "*[t]hey are such as to disclose relationships*" (LW4, 100). It is in this sense that Dewey speaks of scientific activity as a process of meaning-liberation. Dewey is rather explicit on this point:

> [A]s water is taken to be just the thing which we directly experience it to be, we can put it to a few direct uses, such as drinking, washing, etc. Beyond heating it there was little that could be done purposefully to change its properties. When, however, water is treated not as the glistening, rippling object with the variety of qualities that delight the eye, ear and palate, but as something symbolized by H_2O, something from which these qualities are completely absent, it becomes amenable to all sorts of other modes of control and adapted to other uses (LW4, 85).

Scientific activity and scientific concepts are, therefore, first and foremost new and more refined ways of transacting with things in the world. The kinds of activity that are made possible by an experimental setting – the material equipment being understood as part of that activity, in conformity with Dewey's concept of experience – are constitutively different from the ones that are carried out in primary experience. Primary experience consists of the direct manipulation of things present in the proximate environment: these activities rely on the immediate qualities of the things to be handled, and take those quality as reliable grounds for the operations to be performed. So, for instance, the transparency of water is experienced as a relevant quality for drinking purposes. Consequently, the rule of action established by the concept of water will take the following form: if something is liquid and transparent, then if it is drunk it will quench the thirst. A semantic relation is thus established between the quality of being transparent and the fact of quenching the thirst, within the context provided by the life-behavior of drinking and searching for water.

Experimental inquiry, on the contrary, is refined and controlled: as Dewey puts it, it is "*directed activity*", which consists in "doing something which varies the conditions under which objects are directly had and by instituting new arrangements among them" (LW4, 99). The semantic productivity, so to say, of scientific investigations is due to the fact that, through directed activity, the conditions under which an object is commonly used and enjoyed in primary experience are modified so as to enable the production of new consequences and effects.

---

[29]And simultaneously, from that moment on, water – this particular glass of water that I am currently holding – signifies the potentiality to dissolve a certain amount of salt, and not only the potentiality to quench the thirst. The intrinsic relation existing between significance and meaning should always be borne in mind: in conformity with the thesis of the semantic identity of concept and object, the enhancement of the significance of the object goes hand in hand with the enhancement of the meaning and inferential power of the corresponding concept.

Now, the enlargement of the contexts of application of primarily experienced objects goes hand in hand with a remarkable change in the kinds of rules that constitute scientific concepts. Indeed, one of the distinctive features of scientific activity is that it does not take the immediate qualities of the objects as reliable grounds for action. Scientific activity originates from the decision to block the overt activities that primarily experienced objects naturally call out, and to look for other possible connections caused by other, more complex operations. In this context, the transparency of water is not taken as part of the meaning of the concept of water, which establishes a connection between that quality and the operation of drinking, but is treated as a property that has to be explained in and for itself. Such a step back from overt action is achieved by means of what Dewey calls "experimental analysis", which amounts to reducing "objects directly experienced to data" (LW4, 99). These data evoke new operations that bring to light previously undiscovered existential connections, upon which new semantic relations accrue. As a consequence of that, the objects of primary experience acquire new meanings as part of new and more refined contexts of activity.

So, for instance, the transparency of water is discovered to depend on the amount of particles present in it, which, in its turn, is connected to the depth of light penetration. The concept of water – but it would be more correct to say, of H_2O – is thus enriched by the addition of two new rules of action: respectively, if the water is transparent, then if you analyze its chemical composition, you will find relatively few particles; and if the water is transparent, light will penetrate in it more deeply than it would do if the water were less transparent. The latter rule can also be operationalized as follows: if the water is transparent, then if you lower a Secchi disk into it, the Secchi depth will be greater than in the case of less transparent water.[30]

[30]For the sake of completeness, I quote here the whole passage in which Dewey deals with the process of semantic articulation of the concept of water. The quotation is drawn from Chapter 6 of *The Quest for Certainty*, significantly entitled *The Play of Ideas*. "What is wanted is to indicate that once the idea of possible operations, indicated by symbols and performed only by means of symbols, is discovered, the road is opened to operations of ever increasing definiteness and comprehensiveness. Any group of symbolic operations suggests further operations that may be performed. Technical symbols are framed with precisely this end in view. They have three traits that distinguish them from casual terms and ideas. They are selected with a view to designating unambiguously one mode of interaction and one only. They are linked up with symbols of other operations forming a system such that transition is possible with the utmost economy of energy from one to another. And the aim is that these transitions may occur as far as possible in any direction, 1. 'Water' for example suggests an indefinite number of acts; seeing, tasting, drinking, washing, without specification of one in preference to another. It also marks off water from other colourless liquids only in a vague way. 2. At the same time, it is restricted; it does not connect the liquid with solid and gaseous forms, and still less does it indicate operations which link the production of water to other things into which its constituents, oxygen and hydrogen, enter. It is isolated instead of being a transitive concept. 3. The chemical conception, symbolized by H_2O, not only meets these two requirements which "water" fails to meet, but oxygen and hydrogen are in turn connected with the whole system of chemical elements and specified combinations among them in a systematic way. Starting from the elements and the relation defined in H_2O one can, so

I will come back to many of these issues in Chap. 5, where I deal with Dewey's idiosyncratic version of scientific realism. For now what is important to note is that Dewey's pragmatist account of concepts enables him to outline a consistent form of semantic holism, which succeeds in meeting the two demands mentioned above – namely, the view that the meaning of a concept is modified, no matter how relevantly, by adding new information, and that there is a kernel of meaning which remains relatively invariable through the various changes, and which provides the conditions of identity of the concept.

The key move is, clearly, to understand the weak conception of articulation in terms of the idea of enlargement of contexts of practice. A concept can be, and, actually, is, employed in manifold different experiences, and its meaning is subjected to a process of articulation therefor. This implies that the object which is the referent of the concept evokes many new and different operations that, if carried out experimentally, are expected to bring about effects that will confirm the relations between actions and consequences that are established by the rules constituting the concept.

The idea of enlargement of contexts of practice is rather unproblematic. We are all familiar with using an object in different activities or contexts of practices: consider, for instance, a hammer that is used first to drive a nail into a wooden block, and then to block a door from opening. The concept of weak articulation establishes a philosophically interesting relation between the notion of enlargement of contexts of practice and that of semantic enrichment. This conceptual relation is interesting since it makes it possible to accommodate Dewey's insistence on the pluralism of practices within his holistic approach to semantics. Indeed, it is safe to say that the enlargement of the contexts of practice, though semantically productive, is not necessary for gasping the meaning of the "implicit" concept; consistently, its semantic enrichment cannot be necessary either. Intuitively, an agent can 'know' – knowing here means being acquainted, and should not be read as a cognitive activity – how to use the concept of water and how to handle water to quench her thirst, without having any idea of how to use the refined concept of H_2O that states the rules of chemical composition of water. It follows that the layers of meaning that are articulated in and through scientific activities are not necessary to the comprehension and mastery of the corresponding everyday concept. This is a most welcome result since it prevents from adopting a too prescriptive and counterintuitive stance on semantic competence, to the effect that no speaker whoever could be said to master any concept.

At the same time, however, it is evident that when an agent learns how to employ a concept in a new context of practice, she comes to learn a new mode of handling the object that is the referent of the concept. As a consequence of that, the set of rules

to speak, travel through all the whole scope and range of complex and varied phenomena. Thus the scientific conception carries thought and action away from qualities which are finalities as they are found in direct perception and use, to the mode of production of these qualities, and it performs this task in a way which links this mode of generation to a multitude of other" (LW4, 126–127).

that define the structure of the concept becomes more complex and more internally articulated: the concept acquires a new meaning, and the agent who masters that concept acquires the practical ability to respond to the corresponding object in a new and more proper way. So, for instance, the transparency of water, taken as an existing quality, becomes sign not only of its being drinkable, but also of its being composed of relatively few particles; in parallel, the notion of amount of particles becomes part of the meaning of the notion of transparency of water, which is also related with the notion of drinkability.

Accordingly, when semantic holism is formulated in terms of activity and contexts of practice, the idea of meaning articulation can be easily preserved without relapsing into some sort of semantic instability or, even worse, of nihilism about semantic competence. A question remains open, however. It has been repeatedly said that, according to the weak conception of articulation, a concept undergoes a process of semantic enrichment by being employed in different contexts of practice. But how is it possible to say that it is the *same* concept that undergoes a process of semantic articulation? What are the criteria of identity of a concept?

Dewey's solution to this problem is elegant, and, to certain extent, anticipates recent work in philosophy of science in a fascinating way. I have in mind here Chang's reappraisal of operationalism, in the context of his attempt to provide an explanation of the persistence of some epistemic objects despite serious changes in the theories that define them (Chang 2011a). Clearly, Dewey is not interested in the problem of intertheoretical persistence – and not only because he wrote significantly before the publication of Kuhn's *The Structure of Scientific Revolutions*. In his hands, the problem of persistence through theory changes becomes that of understanding the persistence of the same concept in different contexts of practice. Obviously, the most interesting case of semantic persistence is that in which a common-sense concept gets employed and applied in scientific activity. In this sense, there are some points of connection with contemporary approaches.

Now, Dewey's solution revolves around the acknowledgment of the *semantic* primacy of common sense over science. As has been highlighted in the previous chapter, Dewey argues for the genealogical and explanatory primacy of primary experience. Semantic primacy is a slightly different thesis, which consists in identifying the kernel of meaning which warrants the persistence of a concept throughout different contexts of practices with a small set of common-sense operations that any agent is expected to be able to perform. Dewey's formulates this insight with the following words:

> (A) When it was discovered that wood-pulp could be used for making paper if its material was subjected to operations in which it entered into new conditions of interaction, the significance of certain forms of lumber as objects changed. They did not become entirely new substantial objects because old potentialities for consequences remained. But neither was it the same old substance (LW12, 132).
>
> (B) The subjectmatter designated by *water* in popular every day usage is *from the standpoint of what is designated by H_2O* some thing arbitrarily isolated, cut off from its connections [...]. The wisest physicist on earth drinks the temporally-spatially localized water that he sees and touches through immediate use of bodily organs of sight, touch, temperature

> [...]. To the one who is *a*-scientific, as far as use of the procedural method of the latter is concerned, that designated, or stood for, by *water* is unique by virtue of immediate qualities which shut it off from systematic connection with qualitatively unlike things. To one of scientific habit the immediately present [...] is a phase, a specifiable emphasis, of subjectmatter which is indefinitely extensive spatially and enduring temporally (LW16, 306)

From these two quotations taken together, it should not be difficult to grasp what Dewey has in mind. To simplify: take again the concept of water. That concept is originally elaborated and employed in practical life-behaviors such as the search for something to quench the thirst, the act of washing clothes, the irrigation of cultivated fields, and so on. Consequently, the rules that constitute that concept refer to some very basic operations such as "seeing, tasting, drinking, washing," and so on (LW4, 126).[31] In addition, as Dewey explicitly acknowledges, these rules are scattered; they are not integrated into a single coherent system. Nonetheless, the fact that the same object can be used in different contexts of activity – I can drink water and then use the remaining to wash my clothes, for instance – supplies the corresponding concept with a certain degree of unity and coherence.

When water enters into more complex and refined activities, such as a scientific experiment over the chemical composition of liquids, the object acquires new significance and the corresponding concept gets enriched with new meanings. As Dewey states, it is not "the same old substance" anymore since it can be now acted upon in new and different ways (LW12, 132). At the same time, however, it is not an entirely new concept since its new meanings do not erase and replace the primitive ones: no matter how complicated the use of scientific vocabulary and technical equipment may be, the scientist who is capable of handling them has always the possibility to perform the simpler and more basic operations that constitute primary experience. It is precisely because the most basic operations persist and endure – and life-behaviors overlap – that the identification of a concept through different contexts of practice is possible.

So, when the common-sense concept of water evolves into the refined concept of H_2O their (at least partial) identity of reference – the fact that H_2O refers to the same object of water – is warranted by the fact that there is a chain of operations and contexts of practices which makes it possible for the agent to move from one extreme to the other. Let's go back for a moment to the first of Dewey's passages quoted above. I read Dewey as saying that the chain of operations that holds together the various meanings that constitute a concept can be understood both genetically and functionally. On the one hand, that chain has to do with the operations that an agent is supposed to be able to carry out, as already noticed; on the other hand, it has also to do with the history and development of the operations that can be profitably performed on the object.

[31] "How, for example, should the water of direct and familiar acquaintance (as distinct from H_2O of the scientific frame) be described save as that which quenches thirst, cleanses the body and soiled articles, in which one swims, which may drown us, which supports boats, which as rain furthers growth of crops, which in contemporary community life runs machinery, including locomotives, etc.?" (LW16, 245). For an analysis of this passage, see Gautier (2017).

This account, centered around the idea of continuity of operations, brings semantic holism down to earth by focusing on the cultural conditions that lie at the ground of linguistic activity. But those remarks shed also light on a further feature of the weak conception of articulation. As said, one of the theoretical advantages of the latter is that it is not committed to the thesis of the structural instability of meaning, which eventuates in what I have called nihilism about semantic competence. Something more can now be said in this regard. It has been stressed that the insistence on the semantic primacy of common sense concepts enables Dewey to identify the kernel of meaning that allows the re-identification of concepts in different contexts with a set of common-sense operations: the operations that can be carried out on water, as well as its possible conditions of use, can vary indefinitely; nonetheless, water remains that thing that can be used to quench the thirst, wash clothes, irrigate the fields. This is what the idea of the semantic primacy of primary experience amounts to.

Now, in light of what has been just said, it is possible to add that, precisely because it is concerned with the practical affairs of everyday life, primary experience constitutes the common ground in which human beings – no matter how different their scientific education and expertise may be – find a shared semantic frame. It is a fact that those life-behaviors to which all human beings have access, because conditioned and compelled by their biological and cultural endowment, represent the contexts in which an agreement can be easily reached. Primary experience is the semantic bedrock on which all concepts ultimately rest.

Chapter 3
Inquiry

Abstract This chapter is dedicated to the analysis of the temporal pattern of inquiry. This chapter is mainly metaphilosophical: its thesis is that the import of Dewey's idea of inquiry should be 'cashed out' at a metaphilosophical level, in terms of the language that we should use to describe the structure of scientific activities. I argue that Dewey's naturalism, together with his emphasis on the metaphilosophical primacy of the category of activity, puts a number of relevant constraints on what counts as a good philosophical explanation of the different phases that make up scientific inquiries. I offer a detailed reconstruction of Dewey's concepts of situation and tertiary quality, as well as a new account of his notion of warranted assertibility.

Keywords Biological basis of inquiry · Foreshadowing · Serial nature of behavior · Phases of inquiry · Rejection of epistemology · Situation · Normativity · Tertiary quality · Warranted assertibility · Truth · Fallibilism

3.1 Introduction

Dewey's philosophy is grounded in the idea that reality is experienced in many different forms. The manifold ways in which reality shows itself coincide with the plurality of life-behaviors that human beings, as linguistic organisms, can undertake: every life-behavior reveals some aspect of reality, with no single trait of reality being *in se* more fundamental than the others. In some particular cases, the organism experiences a situation as joyful, or boring, or frightening: the elements of the situation are thus enjoyed in their distinctive quality, without paying attention to what they might eventually lead to. In some other cases, however, the situation is experienced – again, qualitatively – as distinctively troublesome; in these cases, reality reveals itself to the organism as incomplete and indeterminate.

Indeterminate situations can be handled in different manners: some of these approaches are instinctive, while some others are reflective and deliberate. When the agent acts instinctively, or she simply relies on previously established habits, she responds immediately to the stimuli of the environment. Her actions may be more or less adequate for satisfying the need that prompted her to act, but in both cases they

R. Gronda, *Dewey's Philosophy of Science*, Synthese Library 421,
https://doi.org/10.1007/978-3-030-37562-1_3

are not an outcome of deliberation. On the contrary, when the agent acts reflectively and deliberately with the aim of solving a problematic situation in a way that could satisfy her demands, pondering the different courses of action that are open to her, she acts rationally. The process of rational resolution of problems is called by Dewey 'inquiry': "*[i]nquiry is the controlled or directed transformation of an indeterminate situation into one that is so determinate in its constituent distinctions and relations as to convert the elements of the original situation into a unified whole*" (LW12, 108). Logic is the name that Dewey gives to the discipline that deals with the structure and nature of inquiry.

Throughout his career, Dewey devotes many efforts to clarifying what inquiry is. The most articulated presentation of his views on the matter is to be found in *Logic, Theory of Inquiry* (1938). However, his reflections on logical issues started many years before the publication of that book. Indeed, already in his early idealistic texts, Dewey confronted himself with the question of understanding the role and function of inquiry in experience.[1] It is Dewey himself who, in the Preface of the *Logic*, stresses the continuity of his reflections on logical subject-matter: he states that there is a thread running through all his texts – *Studies in Logical Theory*, *How We Think* (1910, second revised edition in 1933), *Essays in Experimental Logic* (1916), and finally the *Logic* – and I take it to consist in the idea of inquiry as that moment of experience in which and through which the significances of the objects are revised and reconstructed so as to embody the meanings of the concepts by which they are thought. More precisely, my proposal is to understand inquiry as that process through which the meanings of linguistic signs are controlledly and reflectively transformed into the significances of the objects of the situation.[2]

The goal of this and the next chapter is to lay out, articulate and defend my semantic account of inquiry. And since Dewey analyzes inquiry from two different perspectives, firstly focusing on its general, temporal pattern and then looking for its elements and their interplay, I will follow here the same approach.[3] Consequently, I

---

[1]On the development of Dewey's logical theory, see Johnston (2014). I have analyzed and discussed Johnston's historiographical proposal in Gronda (2015b).

[2]The same point can be formulated in naturalistic terms. From this perspective, the process of inquiry is defined as follows: inquiry is that process through which organisms complex enough to master the use of language change their habits of conduct, and, in so doing, acquire new and better ways of coping with their environment. As a consequence of that, their behavior becomes more attuned to the specific features of the situations they have to face, and purports to be equally successful in dealing with future cases. As Dewey remarks, "[w]hat the organism learns during this process produces new powers that make new demands upon the environment" (LW12, 42).

[3]To my knowledge, it is Brown who first clearly distinguishes between two different dimensions of Dewey's theory of inquiry in Brown (2012, 280). A similar point has been made by Frega, who insists on the distinction that Dewey draws between "distinction between" and "distinction within" in the *Studies in Logical Theory* (Frega 2006, 46). Brown uses the distinction between the two dimensions of inquiry to criticize what may be called the unitary reading of Dewey's theory of inquiry. An example of this kind of reading is Thayer (1952, 50ff.) According to the unitary reading, the antecedent and the consequent conditions of inquiry are to be listed among the phases of inquiry; in so doing, the temporal and the functional accounts of inquiry

will devote the present chapter to outlining Dewey's fundamental notion of situation, while in the next chapter I will tackle his account of the different logical tools – evidence, propositions, judgment – through which the end of inquiry can be reached.

The present chapter is structured as follows. In the first section I will discuss the biological conditions of inquiry – what Dewey calls the biological matrix – for the purpose of highlighting the theoretical import of his idea that reflective activity depends on the biological endowment of the organism. The second section is devoted to laying out what I have called the internal conditions of inquiry. Dewey holds that inquiry is a temporal process whose phases interact one with the others: I will argue that this simple assumption has remarkable metaphilosophical consequences, which can be read as a caveat against any attempt to reduce philosophy of science to epistemology. In the third section I will focus on Dewey's conception of situation, and I will stress its normative ground in the notion of tertiary quality. Finally, in the final section I will discuss Dewey's controversial concept of warranted assertibility, and I will advance a plausible reading of Dewey's puzzling idea that the propositions used as material and procedural means in the process of inquiry do not have truth value.

3.2 The Biological Basis of Inquiry

Inquiry does not come out of nowhere. As a distinctive phase of experience, it has conditions of emergence as well as conditions of fulfillment. Among its conditions of emergence, some of them are related to antecedent phases of experience. Understood in this way, inquiry is the temporal development of those previous phases of experience, which leads to, and eventuates into, a final phase of consummation – "consummation" being a technical term of Dewey's vocabulary which refers to those moments of experience in which the quality of a thing is directly enjoyed in itself, without paying attention to its possible function as sign of something else. Note that the features of this pre-reflective phase of experience put constraints on what counts as a satisfactory conclusion, and consequently on the intermediate phase of inquiry. In this sense, the conditions of emergence of inquiry are internal to experience.

Some other conditions are structural, so to say, and are to be traced back to the biological basis of reflective behavior. Indeed, Dewey argues, inquiry is ultimately made possible by the biological endowment of human organisms. These conditions are not internal to experience, but represent a set of external features without which no inquiry would be possible.

---

are merged together. As Brown has noticed, there are both textual and theoretical reasons to resist this reading. He suggests, therefore, keeping the sequence of inquiry (indeterminate situation-inquiry-reconstructed situation/judgment) well distinguished from the pattern of inquiry (observation-problem-suggestion-reasoning-experiment). I think Brown is right on this point, and I will follow his suggestion here.

The conceptual distinction between external and internal conditions is far from being clear or self-explanatory. As I use it here, that distinction parallels, though in the reverse order, what Dewey says about the biological conditions of experience. As we have seen in Chap. 1, Dewey maintains that the nature of experience has to be clarified by bringing to the fore its structural similarity with the double-barrelled nature of life. In that case, the difference is grounded in the fact that the degree of complexity of experience is higher than that of life, so that any account that overlooks such difference would turn out to be unacceptably reductionist. In this spirit, Dewey notices that any consideration drawn from biological life provides at best a description of the external conditions of experience. Similarly, in the case under discussion here, inquiry is to be conceived of as grounded in specific biological conditions, though such grounding does not entail that the former can be reduced to the latter. I will discuss the first type of conditions – the internal conditions – in the next section. In the present one, on the contrary, I will focus on the external conditions of inquiry.[4]

As Dewey clearly states, the fundamental logical issue is that of understanding "the transformation of animal activities into intelligent behavior having the properties which, when formulated, are *logical* in nature" (LW12, 62). Transformation here is the key term, which is conceptually related to the notion of things passing from "the plane of external pushing and pulling to that of revealing themselves to man, and thereby to themselves" (LW1, 132) – which is the way in which Dewey introduces the notion of meaning in *Experience and Nature*. In both cases, what is at stake is the clarification of the genetic relations between what is pre- (pre-logical, pre-linguistic, and so on) and what is fully and properly articulated.[5] Now, it is evident that the notion of transformation is a form of emergence. The properties of intelligent behavior – which stem from the transformation of animal activities through the medium of language – are original, and cannot be boiled down or reduced to those displayed by animal behavior: they represent an autonomous level of behavioral organization.

Dewey stresses this point also terminologically, by using the term "foreshadowing" to refer to the particular way in which biological activities anticipate and pave

[4]The distinction between external and internal conditions of inquiry is implicitly confirmed by the very structure of Dewey's argument. Indeed, it is not by chance that the first two sections of the *Logic* are dedicated to analyzing what Dewey calls the existential matrixes of inquiry. It is an essential feature of Dewey's naturalism that inquiry, as a complex form of behavior, is rooted in – and emerges from – simpler patterns of agency. From Dewey's naturalistic perspective, inquiry is to be understood as a refinement of the life-behaviors made possible by the biological endowment of an organism complex enough to master the use of language. Since language is, in its turn, the realization of biological potentialities – namely, those potentialities that make it possible for human beings to cooperate in a single course of action – the analysis of the biological endowment of human organisms stands out as the starting point of any sound naturalistic account of inquiry.

[5]Incidentally, it is worth noting that, in the case of logic, the matter is made more complicated by the fact that logical properties are, properly speaking, only implicitly embodied in inquiry, and they are made explicit by a second-level process of reflection which takes as its subject-matter the forms of lower-level inquiries – which are object-oriented inquiries.

the way for reflective activities. He explicitly connects the notion of foreshadowing to the idea of continuity, which in its turn is conceived of in emergentist and anti-reductionist terms.[6] As repeatedly said, I am not interested in discussing the metaphysical implications of this assumption. For my purposes, it is sufficient to provide a general characterization of the most important features of the relation of foreshadowing. A simpler set of activities is said to foreshadow a more internally articulated and complex one when (a) the latter is genetically connected to the former, (b) the simpler set puts constraints on the more complex one, and (c) the more complex set is the realization of some of the potentialities of the simpler set. The last point can be formulated in more neutral – and weaker – terms by saying that there is some sort of dependence between the properties of the complex set of activities and those of the simpler one.

The relation of foreshadowing is, therefore, irreflexive and asymmetric, while it is not clear whether or not it is transitive, though it seems natural to understand it in this way. It also bears explanatory weight: indeed, it is evident that the properties of the simpler set of activities play an explanatory role in accounting for higher-level properties. It is more questionable whether the other direction holds as well. I think it can be argued that higher-level properties are relevant to the explanation of some features of the lower-level properties, but this thesis is far from being evident, and depends on some other metaphysical and conceptual assumptions that I will not discuss here. I will therefore leave the issue open.

With this in mind, we can turn attention to two questions that are worth considering, namely, (a) what are, according to Dewey, the features of animal behaviors that foreshadow intelligent behavior? and (b) what are the consequences, on a theoretical and methodological level, of the emphasis on the biological roots of inquiry?

Let's start with the first issue. Dewey's naturalistic approach moves from the platitudinous assumption that inquirers should be understood as organisms performing a cognitive task. Reflective behavior – which, in its most refined form, is scientific inquiry – ultimately depends on the capacity that some organisms have to withhold action, which is, in its turn, grounded in their capacity to perceive distant objects as possible stimuli for action. There is a qualitative difference, Dewey maintains, between those organisms that lack distance receptors and those which, on the contrary, are endowed with them. Indeed, contrary to the former, the latter are capable of enlarging their environment, thus including spatially and temporally distant events within their course of action.

In order to avoid confusions, notice that Dewey speaks of the enlargement of the *environment* – and not of the *world* – made possible by distance receptors. In Dewey's thought, world and environment are two distinct notions, their distinction

---

[6]Dewey's commitment to emergentism has been acknowledged by many interpreters. See, for instance, Boisvert (1988, 128ff.) and Alexander (1987, 94ff.). See also, for a broader analysis, El-Hani and Pihlström (2002). For a more theoretical approach to the issue, which puts the notion of emergence in relation with the metaphilosophical primacy of the notion of scientific practices, see Pihlström (1999).

being motivated by his willingness to avoid any possible charge of idealism. To simplify: world is a 'neutral', descriptive concept which refers to whatever actually exists; environment, on the contrary, refers to those entities and aspects of the world which matter for an organism. "Matter" here should be understood in the broader sense possible, as encompassing anything that could possibly affect the course of activity of that organism.

Consequently, while there is only one world, there are numerous environments. So, for instance, it can be said that human beings and dogs, while partaking the same part of the world, live in two highly different environments. In this spirit, Dewey remarks that "[t]he environment of an animal that is locomotor differs from that of a sessile plant; that of a jelly fish differs from that of a trout, and the environment of any fish differs from that of a bird" (LW12, 32). But it is also possible to say that different human beings with different training live in different environments: stones have a different significance for a geologist and a kid who plays kicking them around, since they respond to them in different ways.

Let's delve a little bit more into the structure of the relation between organism and environment. First of all, it is clear that is not an *external* relation: organism and environment are essential one to the other. As should be evident, this is a corollary, stated in biological language, of the Semantic Identity Thesis (SIT). "Whatever else organic life is or is not," Dewey remarks, "it is a process of activity that involves an environment". Properly understood, life is a biological transaction that extends "beyond the spatial limits of the organism" (LW12, 32). With this in mind, it should be clear why, according to Dewey, it is wrong to say that the organism lives *in* an environment: a pair of socks is *in* the drawer, since the external conditions do not affect their behavior; on the contrary, the organism lives "by means of an environment", and its life is sustained and fostered by the interaction of intra-organic with extra-organic energies. The transaction between the organism and its environment is, therefore, an act of integration. Or, better said, the organism and the environment "*are* an integration", in the strong sense that the two do not exist one apart from the other (LW12, 32). Whence it is easy to draw the conclusion that "with every differentiation of structure the environment expands" since "a new organ provides a new way of interacting in which things in the world that were previously indifferent enter into life-functions" (LW12, 32).

Dewey's emergentist sympathies lurk behind these statements. His biological naturalism makes room for the possibility of new forms of behavior, which result either from the coming into existence of a new organ or from the modification of the existing endowment of the creature. Now, this thesis has at least two important, strongly interrelated consequences. One is rather obvious, and has to do with the fact that a difference in organs goes hand in hand with a difference in environment. The other one is more interesting, and concerns the need for a *unity* of the environment.

Dewey's argument goes as follows. The differentiation of the organs implies, almost by definition, the differentiation of the processes of integration. At the same time, however, all these different processes of integration are centered on a single organism: they are different ways in which the organism behaves and interacts with its environment. Consistently, the multifarious processes of integration have to be reciprocally integrated, otherwise the organism would be internally disrupted. Dewey formulates this insight with the following words: "[w]ith differentiation of interactions comes the need of maintaining a balance among them" (LW12, 32).

What is relevant for my purposes is Dewey's subsequent remark. Immediately after the quoted sentence, he adds that, "in objective terms", the fact that "[w]ith differentiation of interactions comes the need of maintaining a balance among them" also implies that it comes the need of maintaining "a unified environment". Just as the multifarious processes of integration have to be reciprocally integrated or the organism would be internally disrupted, so the environment by means of which the organism lives is to be unified, otherwise the environment would be internally disrupted too. Dewey is quite explicit in stressing that the two cases are not simply similar; they are identical in all respects since they are to be understood as two ways of describing the very same events. Here is another statement of his position: "[t]he effect of this delicate and complex system of internal changes is the maintenance of a fairly uniform integration with the environment, or – *which amounts to the same thing* – a fairly unified environment" (LW, 33, italics added). This conclusion should come as no surprise in light of SIT.

Now let's shift attention from the general framework to the more technical aspects of Dewey's account of the naturalistic pre-conditions of reflective activities. The capacity to delay action interacts with the existing biological endowment of the creature, thus dramatically changing its modes of response to external stimuli as well as the shape and structure of its environment. So, while in lower organisms the interaction between internal (or organic) and external (or environmental) energies takes place through direct contact, in more complex organisms the modes of response are mediated by and through distance receptors and special organs of locomotion (LW12, 35). Locomotion and distance receptors put the organism in contact with distant parts of the world that had previously no relevance for it; in doing so, they convert them in potentially significant elements of its environment. Because of the new kind of integrations made possible by distance receptors and organs of locomotion, the behavior of the organism becomes serial: the organism does not merely respond to the external stimuli that continuously push on its surface, but organizes its response in light of the ends that it wants to achieve. The modification of the spatial relations among the elements of the environment entails, therefore, the accrual of temporal relations on the experienced material.[7]

Dewey formulates this point with great clarity: "the serial nature of life behavior demands that earlier acts in the series be such as to prepare the way for the latter" (LW12, 35). In lower organisms, the interaction by direct contact shortens the time

[7]It goes without saying that Dewey is not making a case for the reduction of temporal qualities to spatial ones; neither is he attempting to establish a logical connection between the two. The argument puts forth in the second chapter of the *Logic* is directed to highlight the internal relation existing between them in the context of the organism-environment integration that foreshadows linguistic experience. The thesis is therefore not metaphysical, but is an analysis of the biological conditions by which spatial and temporal relations become existentially relevant (directly or indirectly) to an organism.

between the occurrence of the stimulus and that of the response: accordingly, the latter follows almost immediately to the action of the former. On the contrary, in more complex organisms the response takes more time to be performed. If I see a car running fast towards me, because of my distance receptors the car is a significant part of my environment long before it eventually passes nearby. Consequently, the consequences of my future transactions with the car become a factor influencing my present behavior, and affect my decisions about which action to undertake. Similarly, since I have the capacity to move, the search for a distant object may easily become the end-in-view of a course of action. Even in this case, some possible future effect directs and influences the acts that I am interested in performing now, and puts them in a temporal series controlled by a distant (both spatially and temporally) goal.

All these considerations should be read in light of Dewey's distinction between the notion of organic circuit and that of reflex arc.[8] In his groundbreaking article *The Reflex Arc Concept in Psychology* (1896), Dewey stresses that human behavior cannot be properly accounted for in terms of reflex-arc relationships between stimuli and responses. The paradigmatic example of reflex-arc relationship is the knee-jerk reaction. A certain stimulus is necessarily followed by a certain response, 'necessarily' here meaning that no deliberation can occur between the former and the latter. Indeed, the distinctive trait of that kind of reactions is that they are mechanical and not susceptible of correction or further adaptation. The stimulus of a reflex-arc reaction cannot modify the response, and *viceversa*. In other words, the two elements are external one to the other, as a consequence of which no process of revision and modification of their relation is possible. To understand human behavior in terms of reflex-arc relationships implies, therefore, the mechanization and parcelization of human activity: human life-behavior becomes a series of disconnected mechanical reactions with no possible reciprocal *internal* interaction, "a patchwork of disjointed parts, a mechanical conjunction of unallied processes" (EW5, 97).

On the contrary, the idea of organic circuit points into the opposite direction. "What is needed", Dewey remarks, "is that the principle underlying the idea of the reflex arc as the fundamental psychical unity shall react into and determine the values of its constitutive factors" (EW5, 97). Stimuli and responses are to be understood as mutually defined, and such mutual definition entails that we should stop thinking of them as two separate, individual entities – first comes the stimulus and then, after a while, the response follows – and start conceiving them as two moments of a broader continuum. This is a standard Deweyan move, which amounts to shifting the focus of attention from the parts to the whole – the sensori-motor circuit – within which they are *constituted* as elements. A stimulus for action gets

[8]Much has been written on Dewey's concept of reflex arc. For an accurate analysis of this notion, see Tiles (1990, 45ff.). More recently, Santarelli has provided a detailed reconstruction of the various explanatory functions that the notion of reflex arc performs in Dewey's thought (Santarelli 2016).

defined only when a certain train of response has started: before that, its significance is completely undetermined *as a stimulus*. Here is what Dewey writes in this regard: "sensation as stimulus does not mean any particular psychical *existence*"; rather "[i]t means simply a function, and will have its value shift according to the special work requiring to be done" (EW5 107).[9]

What is primary is, therefore, the concrete activity in which the organism is engaged. So, in the familiar case of the child burnt by a flame, what is primary is the act of seeing, while the visual sensation of a candle or that of a light come second: "if this act, the seeing, stimulates another act," Dewey writes, "the reaching, it is because both of these acts fall within a larger coordination" (EW5, 98). This thesis is philosophically relevant. Among other things, it means that it is not properly correct to understand the response as a response *to* a stimulus; rather the contrary, it should be said that the response is *into* the stimulus. "The burn", Dewey continues, "is the original seeing, the original optical-ocular experience enlarged and transformed in its values"; consistently, "[i]t is no longer mere seeing; it is seeing-of-a-light-that-means-pain-when-contact-occurs" (EW5, 98).

It follows that when the larger coordination is taken as primary, the relation between (the function of) stimuli and responses turns out to be *internal*. Now, a relation is internal – in this technical sense – when a process of coordination takes place. While in the case of reflex-arc relationship what comes before causes the existence of what comes after, in a process of *co*-ordination things are much more complicated. Not only what comes before causes what comes after; it also affects its significance. And, at the same time, what comes after is able to influence the significance – not the existence, since that would be a sort of mysterious action – of what comes before. As Dewey remarks, "[t]he stimulus is simply the earlier part of the total coordinated serial behavior and the response the later part" (LW12, 37).[10]

According to the organic circuit model of activity, therefore, the structure of serial behavior is not a temporal sequence of disconnected acts, but a continuous flow in which the present acts are such to prepare the way for the future ones – both in terms of their existence and of their significance – and, at the same time, the latter modify the significance of the former. In complex organisms the activities aroused by distance receptors enter into conflict with contact-activities, and the tension that follows from that conflict is solved by distance-activities taking over and controlling the course of action. Thus, the immediate responses to present stimuli

[9]Dewey's insistence on the difference between existence and function is due to the fact that the quotation is drawn from the early essay *The Reflex Arc Concept in Psychology*; however, the very same idea can be easily accommodated within his mature naturalistic framework, and formulated in terms of the different possible forms of behavioral response to environmental stimuli (see above, Sect. 2.2).

[10]In the *Logic* Dewey reformulates this insight in terms of the distinction between stimulus-response, on the one hand, and excitation-reaction, on the other. Here, the couple excitation-reaction refers to the mechanical response to an organic state of imbalance, and its distinctive feature is that of being isolated and complete in itself. On the contrary, the couple stimulus-response indicates the complete organic circuit of reciprocal co-determination (LW12, 36).

are suspended, and the different activities in which the organism is progressively engaged are determined and evaluated in light of the goal to be reached. As a consequence of the suspension of the immediate discharge of energies, the behavior of the organism acquires a serial structure.

The suspension of the immediate discharge of energies made possible by the endowment of complex organisms is the biological basis of inquiry. Without it, the capacity to figure out and evaluate different courses of action would not be possible. When an organism feels a state of imbalance – a state of objective need – the organic tendency to get rid of it in the simplest and most immediate way is thus blocked, and is coordinated with many other organic processes. It is this totality of coordinated organic activities that, properly speaking, constitutes the stimulus for action (LW12, 36). Being the stimulus "the tension in the total organic activity" which stems from the contrast between contact-activities and those that are caused by distance receptors, its effect as stimulus is not immediate, but persists throughout the entire activity. Dewey formulates this point in the following way: "[t]he so-called stimulus, being the total state of the organism, moves of itself, because of the tensions contained, into those activities [...] which are called the response" (LW12, 36–37). Clearly, since this process takes time, with the responses affecting the significance of the stimulus, the latter "changes in its content at each stage of the [process]", while persisting through it. As will be made clear below, this biological feature of human behavior has extremely remarkable consequences at the level of reflective activity.

I do not want to tackle the issue of the scientific validity of Dewey's statements about the biological constitution of complex organisms; these are empirical theses, and, as such, they are to be empirically tested and investigated.[11] What I want to highlight, rather, are the philosophical consequences that Dewey cashes out from those empirical theses. In particular, I will focus on the theoretical implications that follow from the acknowledgment that inquiry is a biological phenomenon – and, more precisely, on their broader meta-philosophical import on what counts as a good philosophical explanation of scientific inquiry.

As said, inquiry is a serial behavior, which is grounded in the organic circuit of stimulus and response. It is, therefore, a temporal activity, in a threefold sense. First of all, it is a temporal activity in the trivial sense that it takes time to be conducted and brought to an end. Secondly, it is a temporal activity in the theoretically more relevant sense that the elements through which inquiry is carried out – which Dewey calls *logical* elements[12] – have an essentially temporal character. As will be shown in the next section, Dewey's emphasis on the temporal quality of reflective behavior

[11]In recent times, an attempt has been made to translate many Deweyan – and, more in general, pragmatist – theses into the language of contemporary cognitive sciences and neurosciences. Among the most notable examples of this approach, see Solymosi and Shook (2014), Shook and Solymosi (2014), Schulkin (2015), and Madzia and Jung (2016).

[12]A terminological remark is in order here. In Dewey's vocabulary, 'logical' is used to refer to the property that a certain element possesses in so far as it has the capacity to contribute to the successful solution of the problem that prompted the inquiry. It has nothing to do with the standard,

or inquiry is intended to pave the way for the realization that the logical forms and the concepts deployed in the course of inquiry are intermediate rather than final. Their function is that of preparing the solution of a problem: they are tools for intelligently directing the inquiry, and, consequently, for deliberately transforming the existential conditions that called out reflection. From this perspective, they are homologous to the intermediate responses that, within a course of biological activity, lead to the attainment of an end.

Thirdly, and most importantly, inquiry is a temporal activity in the strong sense that it entails a modification of the environment – *any* inquiry changes once for all the objective conditions that the agent of inquiry has to face. This point is closely related to the one just made. It is precisely because any activity put forth by an organism dramatically and irreversibly changes the environmental conditions that the capacity to delay and control action is so important from an evolutionary point of view. Such capacity makes it possible for the organism to stop from acting in a possibly harmful way, and enables it to plan some possible courses of action that are better fitted to deal with the problematic situation.

This point can be easily seen on a biological level. Consider, for instance, a complex organism, endowed with distance receptors, that is craving for food. From a naturalistic perspective, hunger is a state of imbalance between the needs of the organism and the environmental conditions with which it is confronted. The organism is compelled by such state of imbalance to move and search for a prey. Now, it is evident that, in order to reach that goal, it has to undertake some course of action. It is also evident that not every course of action is equally promising, and many of them are, on the contrary, potentially harmful. Being irreversible, once a modification of the environmental conditions is brought about, its consequences are beyond the control of the organism that prompted them. So, its various acts are to be coordinated in such a way that the existential modifications that each one of them provokes succeeds in promoting the attainment of the end-in-view.

Let's move now from the biological level to the level of inquiry. The key notion is that of existential modifications. Any inquiry, no matter how abstract and theoretical it may be, modifies the objective, existential conditions of the situation: even an astronomical observation, Dewey remarks, ultimately depends on some existential modification of the experimental setup – for instance, the modification of the position of the telescope (LW4, 68).[13] These transformations are necessary in order

contemporary meaning of the term. In this text, 'logical' is used in the Deweyan sense: a logical element is anything that performs a function within inquiry.

[13]"In astronomy, for example, we cannot introduce variation into remote heavenly bodies. But we can deliberately alter the conditions under which we observe them, which is the same thing in principle of logical procedure. By special instruments, the use of lens and prism, by telescopes, spectroscopes, interferometers, etc., we modify observed data. Observations are taken from widely different points in space and at successive times. By such means interconnected variations are observed. In physical and chemical matters closer at hand and capable of more direct manipulation, changes introduced affect the things under inquiry. Appliances and re-agents for bringing about variations in the things studied are employed. The progress of inquiry is identical with advance

to change the objective conditions in a way that make it easier for the agent to come up with a solution to the problem that originated the inquiry. The parallel with the biological description of the behavior of an organism searching for food is evident.

In Dewey's hands, such a parallel between biological and reflective behavior is theoretically fruitful. Two consequences are particularly relevant. First of all, the overt activities put forth by the hungry organism lead to a series of interrelated existential modifications that eventuate – if they are adequate as means to reach the end-in-view – in the actual enjoyment of the caught prey. Similarly, the overt activities put forth by the agent in an experimental context lead to a series of interrelated existential modifications that eventuate – if they are adequate as means to reach the end-in-view – in the actual enjoyment of the reconstructed situation: in the case of the astronomical observation, for instance, the discovery of the trajectory of the observed planet. The outcome of any inquiry is the result of a series of interconnected existential modifications. Just as the hungry organism cannot satisfy its hunger under present environmental conditions, so the inquirer cannot get on top of the problem by accepting and relying on the existent experimental conditions. She has to act in a controlled and deliberate way for the purpose of existentially changing those aspects of the situation that are perceived as problematic.

Secondly, the structural similarity between biological activity and logical inquiry lies at the core of Dewey's *epistemological* anti-representationalism. It is worth noting that, though the two are intrinsically related, Dewey's *epistemological* anti-representationalism has to be distinguished from his *semantic* anti-representationalism.[14] The main features of the latter have been discussed in the previous chapter so I will not dwell on it any further here; the former consists in the thesis that the outcome of inquiry does not aim to faithfully represent things as they are in themselves: indeed, the experimental conditions that – according to the representationalist theory of knowledge that Dewey rejects – inquiry should represent are precisely those conditions that inquiry is supposed to existentially change.

This point can be formulated in the clearest and most straightforward way by referring to Dewey's definition of inquiry as the transformation of an indeterminate or problematic situation into a determinate one. As has been observed, an indeterminate situation is characterized by the fact that the significances of some of its elements are not clear, either because they are not clear in themselves or because the significances of some elements enter into contrast with those of others. Consequently, the representation of the existing significances of the elements of the situation is completely ineffective from a practical point of view, i.e., from the point of view of the course of action that has to be undertaken by the inquirer. Put roughly,

---

in the invention and construction of physical instrumentalities for producing, registering and measuring changes." (LW4, 68).

[14]For an accurate analysis of Dewey's epistemological anti-representationalism, which focuses on Dewey's theory of representation, as well as on his rejection of the idea that knowledge is representation, see Godfrey-Smith (2019).

a problematic situation is a mess, which blocks the habitual ways in which the agent responds to the environmental stimuli, and calls for reflection. In this context, it is clear that it makes no sense to argue that the goal of inquiry is that of depicting a mess. The adequate and reliable representation of the existing conditions is clearly an indispensable tool for a successful inquiry, but it cannot be its end.

Dewey's insistence on the objective and existential character of the modifications brought about by inquiry has therefore to be seen as part of a broader and more ambitious metaphilosophical project concerning the language in which cognitive activities can be adequately described and analyze. Dewey acknowledges that the adoption of a language that emphasizes the continuity between lower biological activities and higher linguistic reflective behaviors is functional to the development of a naturalistic theory of inquiry that challenges the legitimacy of epistemological analysis. Indeed, to say that inquiry should be understood as a process of objective modifications of the experimental environment implies that it cannot be reduced to a series of 'private', mental episodes – in contemporary terminology, a series of states of belief – that convey information about some worldly state of affairs.[15]

So, instead of postulating faculties or private states of belief, Dewey argues that the cognitive episodes that constitute inquiry are, first and foremost, publicly and experimentally detectable events, which should be explained in terms of overt behavior.[16] Notoriously, this view is formulated by Dewey in terms of an adverbial

[15]As should be evident, this view is intrinsically related to Dewey's externalism, to the effect that the content of a mental state is determined by its being part of the situation in which the agent having that state happens to find herself involved. It is with this idea in mind, for instance, that it is easy to draw a distinction between a genuine state of happiness, which is causally related and dependent on the objective features of the environment, and a pharmacologically induced state of euphoria which is phenomenologically indistinguishable from the former (see, in this regard, LW10, 48). And it should not be difficult to see that that conclusion follows directly from the thesis that an organism is not, properly speaking, *in* an environment, but lives through and by the environment which supports its activities. It is the transactional whole within which the organism and its environment are included that provides the conditions for the fixation of the content of its mental states.

[16]This aspect of Dewey's philosophy of science has been lucidly pointed out by Levi, who has also stressed the difference between Peirce and Dewey's conceptions of inquiry, and has openly argued for the theoretical superiority of the former over the latter. This should come as no surprise. Indeed, while Peirce's theory of inquiry can be easily formulated within the framework of Levi's logic of belief-revision, such translation is not possible in the case of Dewey's account: "Inquiry according to Dewey 'is the controlled or directed transformation of an indeterminate situation into one that is so determinate in its constituent distinctions and relations as to convert the elements of the original situation into a unified whole.' I prefer Peirce's assertion that the aim of inquiry is the removal of doubt. This is not merely predilection for one style of formulation over another. Peirce's characterization can readily be rephrased as involving a transformation of an initial state of doubt into a state in which the doubt is removed. This suggests that the transformation is of one state of belief by another (or more generally of one point of view by another if it is important to take into account attitudes other than full belief such as states of probability judgment and value judgment). Dewey explicitly resisted formulations of this kind" (Levi 2010, 83). Levi is certainly right on this point: it is beyond doubt that Dewey's approach is not easily and immediately translatable into the language of belief-revision theory. Nonetheless, I think that it is not impossible to make

theory of cognition: since every cognitive capacity of human organism (mind, rationality), as well as every property of a cognitive state (justification, warranted assertibility), is a feature of a specific kind of activity, it can be traced back to a particular set of empirical detectable qualities of human behavior. So, for instance, Dewey suggests not talking of reason or rationality, and using rather the expression "acting rationally". Similarly, instead of speaking of mind, he recommends to use the adverb "mindfully", which describes some distinctive features of a certain kind of behavior, such as the quality of paying attention to the consequences of the acts to be performed or that of being careful in searching for the data of the problem.[17]

These metaphilosophical constraints should be met by any consistent theory of inquiry. Accordingly, they set the stage for Dewey's analysis of the sequence of inquiry (indeterminate situation- inquiry- determinate situation) by establishing which conceptual tools can be deployed to account for the distinctive features of knowing and scientific activities. In the following sections these metaphilosophical constraints will be put at work and further articulated.

3.3 The Internal Conditions of Inquiry

The previous section was devoted to analyzing the external conditions of inquiry. In the present section, on the contrary, I will turn attention to what I have called its internal conditions. "Internal conditions of inquiry" is not a Deweyan formula. Dewey speaks of "antecedent conditions" of inquiry, but the two notions – though strictly connected – are not identical. Broadly speaking, the notion of internal

---

room for the logic of belief-revision within Dewey's philosophical framework. The key move here is to understand the language of doubt and belief – the one that is needed in order to formulate the concepts of 'state of full belief', 'doxastic commitment' or 'belief-change' which lie at the ground of Levi's analysis of the process of belief-revision – as a kind of abstraction from a more fundamental phenomenon, namely, inquiry as the controlled transformation of an indeterminate situation into a determinate one. Clearly, Dewey is suspicious of *philosophical* abstraction, but is *not* suspicious of philosophical *abstraction*. A context has to be provided that makes it clear that, when using the language of doubt and belief, we are handling a set of highly refined conceptual tools.

[17]See, for instance, what Dewey writes in *Experience and Nature*: "'Thought', reason, intelligence, whatever word we choose to use, is existentially an adjective (or better an adverb), not a noun. It is a disposition of activity, a quality of that conduct which foresees consequences of existing events, and which uses what is foreseen as a plan and method of administering affairs" (LW1, 126). A similar point is formulated in *Reconstruction of Philosophy* in regard to the concept of truth: "The adverb 'truly' is more fundamental than either the adjective, true, or the noun, truth. An adverb expresses a way, a mode of acting. Now an idea or conception is a claim or injunction or plan to *act* in a certain way as the way to arrive at the clearing up of a specific situation. When the claim or pretension or plan is acted upon *it guides us truly or falsely*; it leads us to our end or away from it. Its active, dynamic function is the all-important thing about it, and in the quality of activity induced by it lies all its truth and falsity. The hypothesis that works is the *true* one; and *truth* is an abstract noun applied to the collection of cases, actual, foreseen and desired, that receive confirmation in their works and consequences" (MW12, 169–170).

conditions is theoretically thicker than Dewey's concept of antecedent conditions in that it purports to pick out those general, structural features of experience that make inquiry possible. The rationale behind that notion is that any phase of experience that affects another phase counts as one of its internal conditions. The antecedent conditions of inquiry are, therefore, part of its internal conditions since they are, by definition, those phases of experience from which reflective experience arises. But the consequent conditions of inquiry counts as internal conditions too, since they provide the ultimate criteria of evaluation of the satisfactoriness of a course of inquiry.

In addition to this, however, it might be argued that the particular kind of normativity that holds together the different phases of experience should also count as an internal condition of inquiry: indeed, the latter would not be possible without some sort of connection among the antecedent conditions of inquiry, the course of inquiry itself, and the consequent condition. This is certainly correct, but since normativity is not, properly speaking, a phase of inquiry – at least, as Dewey understands it – but rather a quality of the whole course of activity, I will not include it among the internal conditions. Consequently, I will defer its discussion to the next section and restrict the notion of internal conditions to refer to those phases of experience which are not cognitive, and which nonetheless are necessary in order for inquiry to occur. In so doing, I follow closely what Dewey writes in the first essay of the *Studies in Logical Theory*, in which he acknowledges that "the very heart of the logical problem" is "the relation of thought to its empirical antecedents and to its consequent" (MW2, 298).[18]

As I read it, the idea of internal conditions of inquiry – that is, the fact that inquiry does not come out of nothing and it is not and end in itself, but stems from, and issues in, a phase of experience which is not cognitive – has relevant consequences mainly at the metaphilosophical level, and, consequently, is part of the rich metaphilosophical apparatus through which Dewey marks his distance from the traditional account of knowledge and scientific activity. In particular, I believe that Dewey's emphasis on the derivative and secondary character of inquiry is functional to marking the distinction between his practical conception of the knowing activity

[18]For the sake of completeness, I report here the whole quotation: "[n]o one doubts that thought, at least reflective, as distinct from what is sometimes called constitutive, thought, is derivative and secondary. It comes after something and out of something, and for the sake of something. No one doubts that the thinking of everyday practical life and of science is of this reflective type. We think about; we reflect over. If we ask what it is which is primary and radical to thought; if we ask what is the final objective for the sake of which thought intervenes; if we ask in what sense we are to understand thought as a derived procedure, we are plunging ourselves into the very heart of the logical problem: the relation of thought to its empirical antecedents and to its consequent, truth, and the relation of truth to reality". A point is worth making. The quotation may be read as implying that the logical problem consists of three different issues, (a) the relation of thought to its empirical antecedents and consequent, (b) truth, and (c) the relation between truth and reality. As I read Dewey's position, this interpretation is incorrect. I suggest seeing the three questions as three different yet converging ways of formulating and understanding the very same point, i.e., the intermediate character of knowledge.

and two other accounts of knowledge – one idealistic and the other epistemological – that he deems as fatally flawed. I will, therefore, present Dewey's views while reviewing the criticisms he directs to the positions he rejects. Those accounts are unsatisfactory in that – no matter how different they might appear – they share a common commitment to the representational view of knowledge. In both cases, indeed, the criterion of validity of knowledge is traced back to the achievement of an adequate representation of something which exists before and independently of knowing activity: according to the idealistic account, this 'something' is the activity of constitutive thought; according to the epistemological account, it is an independent reality which our thoughts purport to represent (LW4, 87).

Let's start, very briefly, with the idealistic account. As is well know, Dewey was reared in the tradition of Anglo-American idealism: his first works stem from the debates concerning the relationship between psychology and philosophy, and his early theory of logic is one of the ways in which Dewey tries to address that problem. No surprise, therefore, that in his early texts he devotes a great deal of attention to stressing his disagreement with other contemporary idealistic approaches. For the very same reason, almost no trace of that debate can be found in his later works: once the naturalistic standpoint has been conclusively articulated and secured, Dewey has no reason to deal with that theoretical option anymore.

In the *Studies in Logical Theory*, Dewey's point of disagreement with the idealistic tradition is expressed with the greatest clarity. There is no doubt, Dewey writes, that "thought, at least reflective, as distinct from what is sometimes called constitutive, thought, is derivative and secondary" (MW2, 298). Dewey's analysis is couched in the language of experience, which makes things more complicated to grasp. Nonetheless, the point that he is trying to make is rather clear: if one pays attention to what one does when she reflects upon a certain problem, it is possible to recognize "a certain rhythm of direct practice and derived theory" (MW2, 299). First comes the direct activity in which the agent is engaged, and then follows the activity of reflection through which the objects encountered are reflected upon.

This is what the naive standpoint – i.e., the standpoint of an agent who is not interested in the philosophical problems that the activity of reflection may originate – tells about inquiry. As an activity, inquiry has specific conditions of occurrence, as well as specific conditions of satisfaction or success: inquiry arises when some direct course of activity is blocked and interrupted, and new ways of dealing with present existential conditions are to be developed. It is for this reason that Dewey calls reflective thought derivative: it is derivative in the strong sense that it cannot create its own material, but has to draw it from the other activities in which the agent is engaged. In addition, being derivative, it cannot purport to be an end in itself since inquiry is undertaken for the sake of direct activity. The function of inquiry is that of providing the agent with new and more effective meanings – in naturalistic terms, habits of action – through which she can handle the problem that called up reflection in a controlled and satisfactory manner.

The idealistic account of thinking activity, on the contrary, rejects the naive perspective on inquiry, and tries to address a wholly different question. According to the idealistic approach, indeed, the problem with which philosophy is concerned is not that of understanding how a particular agent can undertake a particular inquiry under a set of particular conditions. This is an empirical fact that has no real philosophical value. The problem is rather that of explaining how it is possible for the world to be experienced as significant, and, consequently, of understanding what role is left for concrete, reflective processes of inquiry.[19]

Now, Dewey rejects the idealistic account of thought as explanatory harmful. Firstly, it makes it difficult to explain, on *logical* terms, why there should be any need of reflection if the work of production of meaning has been secured once for all by the constitutive activity of thought. Secondly, it leads to unnecessary metaphysical problems since, by presupposing the existence of something – a thing, a faculty, or a function – which cannot be, in principle, described in empirical terms and yet must be somehow accessible to the empirical agent, it creates the problem of understanding how that relation is possible.

I do not want to assess the soundness of Dewey's criticisms, since that is an issue which has now a merely historiographical interest. The point that I would like to make is more general, and has to do with the ultimate reason of Dewey's dissatisfaction towards the idealistic account of inquiry. As said above, I think that the disagreement between the two approaches is metaphilosophical rather than

[19]The idealistic argument, at least as Dewey presents it, goes as follows. It is made up of two premises. The first one is a strong philosophical assumption: idealists hold that the activity of thought is a necessary condition for meaning – meaning in general, without further qualification – to be given. Clearly, this is a controversial assumption, but it has some *prima facie* plausibility. At the end of the day, it is not *a priori* implausible that meaning ultimately depends on our capacity of drawing connections between different aspects of different objects. The second premise is uncontroversial: it is a fact that the greatest part of the meanings with which we are acquainted is given unreflectively – i.e., it is given previously and independently of an act of reflection. Furthermore, this premise is also a necessary presupposition of Dewey's theory of inquiry as an intermediate and derivative phase of experience. In Dewey's terminology, objects are experienced as *significant* within the context of a certain direct activity, and inquiry arises only in those cases in which the significances of the objects of a situation – which is now experienced as indeterminate – are either too vague or in contradiction one with the others. Stated in more technical terms, both the antecedent and consequent conditions of inquiry are phases of direct experience, that is, phases in which the objects that are experienced are immediately used and enjoyed. In those phases, the significances of objects are given, which means that habits of behavior are settled and continuously confirmed and supported by the responses of the environment. Or, put in semantic terms, this means that in those phases meanings and significances are well attuned. Inquiry, Dewey remarks, takes "some part of the universe of action, of affection, of social construction, under its special charge, and having busied itself therewith sufficiently to meet the special difficulty presented, thought releases that topic and enters upon further more direct experience" (MW2, 299). Accordingly, Dewey cannot but accept this premise of the argument. It follows from those two assumptions that the activity of thought that should account for the immediately experienced significances cannot be identified with deliberate reflection. From this consequence idealists draw the conclusion that there must be some sort of constitutive activity of thought which performs its function behind the curtain of empirical consciousness, so to say. The goal of reflective activity becomes that of reproducing the processes of the original, constitutive activity – a result that can be achieved by pruning off the limitations due to the empirical character of the inquirer. Incidentally, it is worth noting that the distinction between significance and meaning is, among the other things, a conceptual move that precludes the very possibility of formulating the idealistic argument.

properly philosophical. Dewey believes that the idealists – what Dewey has in mind here is Thomas Hill Green's account of the Absolute Self, but his remarks can be easily extended to other versions of idealism – failed to realize that "[t]he substitute for this objective constitutive thought process [...] is the practical or teleological organization, organization through our past habits" (DL, Vol. 1, Part II, 625). In doing so, by extending beyond its empirically detectable limits the notion of reflection – an extension which follows directly from the idealistic tenet that there can be no meaning without the synthetic activity of thought – they have ended up overintellectualizing direct experience, thus blurring the distinction between a kind of activity that has to be accounted in purely biological terms and an activity that has to be formulated in the language of rational discourse.

The same kind of metaphilosophical disagreement lies at the basis of Dewey's rejection of the project of epistemology.[20] Indeed, even in this case the difference between Dewey's approach and that of his rivals can be formulated as a disagreement over the scope of applicability of a specific kind of philosophical language. Be it the language of epistemology or the language in which the distinction between constitutive and reflective thought can be drawn, the idea that Dewey wants to convey is that the language used by philosophers to describe the episodes of knowing may not be attuned to the goal.

An example can be useful here to fix the point. As is well known, epistemology is concerned with the study of knowledge. One of the tasks that epistemology is asked to accomplish is to provide a satisfactory analysis of the different sources of knowledge: in order for something to count as knowledge, indeed, it has to originate in a source or faculty that we have good reasons to consider reliable. Traditionally, perception is taken as one of those sources, and perceptual knowledge is usually conceived of as the paradigmatic form of reliable knowledge – consider, for instance, the fact that a consistent number of Gettier cases have to do, and are built upon, instances of perceptual knowledge.

Now, Dewey rejects the very idea that perception is, *in itself*, a case of knowledge; in so doing, he also rejects the metaphilosophical assumption that the language of epistemology has unlimited applicability and validity. Dewey's point is theoretical sophisticated. According to Dewey, far from being a case of knowledge, perception is a biological phenomenon that has nothing to do with knowing activity. From Dewey's naturalistic perspective, perception is rather to be understood as the *effect*

[20]Dewey's criticism of what he calls the "epistemological industry" has drawn the attention of numerous scholars (MW1, 122). Dewey's practical turn, together with his insistence on the fact that knowledge is an activity rather than a mental state, has been acknowledged as the ultimate source of his rejection of the epistemological program. For a clear exposition of the main features of Dewey's theory of knowledge, see Putnam (2010). A defense of Dewey's attempt to substitute epistemology with the theory of inquiry can be found in Thayer (1990) and in Hickman (2007). For an analysis of Dewey's fundamental notion of judgment of practice, in which the practice-oriented approach is used to develop a consistent naturalistic 'epistemology' (in the pragmatist sense) that aims to reconcile moral and scientific reasoning, see Welchman (2002) and Frega (2010, 2012). For a broader analysis of the pragmatist approach to epistemology, see the collection of essays in Frega (2011).

of the biological transaction between the organism and its environment. More precisely – and I quote here from Dewey's *The Need for a Recovery of Philosophy* (1917) – "it is an affair of the dynamic interaction of two physical agents in producing a third thing, an effect"; as a natural event, it is "an affair of precisely the same kind as in any physical conjoint action, say the operation of hydrogen and oxygen in producing water" (MW10, 31). It is only when the naturalistic standpoint is abandoned in favor of the epistemological stance that a biological event – which should be investigated either in its physiological bases or in its ecological conditions – becomes a cognitive phenomenon, and the issue of assessing its reliability as a source of knowledge becomes a philosophically relevant problem.[21] Change the language of description; the epistemological problem dissolves.

Dewey's favorite example – which is now quite outdated, but which was important for the epistemological debate of his time, centered upon the contrast between representative or direct realism[22] – is the spherical object that presents itself to one observer as a circle and to another as an elliptical surface. Since there is only one real object, so the realist argument goes, the differences in the two perceptions must depend on the distorting influence of the two subjects. Dewey's countermove is to argue that all these epistemological problems disappear when the facts in the question are translated into naturalistic terms. From a naturalistic perspective, indeed, there is nothing strange or surprising in the fact that light is refracted in different ways, depending on its angle of incidence. As Dewey remarks, the opposite would be true: "[i]f the same spherical form upon refracting light to physical objects in two quite different positions produced the same geometric forms, there would, indeed, be something to marvel at" (MW10, 31).

Dewey's point is that, in this as in many other cases, there is no use for an epistemological approach that, far from having any explanatory value, creates philosophical problems where there are none. The two different perceptions of an elliptical or of a circular body are not cases of knowledge; as Dewey explicitly states, "[t]hese forms are natural happenings", and the only interesting explanation that is needed is the one provided by natural sciences (MW10, 32). Clearly, those natural phenomena may easily become objects of knowledge; for instance, they can be made the subject-matter of physical or psychological investigations. Nonetheless, they are not cognitive *in themselves*, as epistemologists hold. On the contrary, they acquire cognitive value only if and when they are *reflected upon*.

Now, it is only when natural happenings are reflected upon, thus becoming the subject-matter of a reflective activity, that the operations through which they are

[21]See in particular for this kind of metaphilosophical approach to the problem of perception Dewey's essay *The Naturalistic Theory of Perception by Senses* (LW2, 44–54), as well as Dewey's criticisms to Russell's *Our Knowledge of External World*, formulated in *The Existence of the World as a Logical Problem* (MW8, 83–97).

[22]For a broad-brush description of the debates between pragmatists and realists, see Gronda (2018a, 290–94). For a more detailed reconstruction of Dewey's relationships with his contemporary realists, see Hildebrand (2003). See also Shook (1995) for an accurate account of Dewey's confrontation with the American realists in the first decade of the twentieth century.

investigated are controlled and checked in light of the aim that they are expected to satisfy. Since these are deliberate and controlled activities, they can be profitably described in and by a language centered on the notions of truth, falsity, reliability, and so on. Indeed, in this case, the agent can adopt a critical, reflective stance on her own activities, thanks to which her epistemic performances may increase in quality and efficacy. In this case, consequently, the epistemological language is not idle – it does not spin in a frictionless void. Rather the contrary, it makes a concrete difference in the behavior of the agent.

The traditional epistemological approach, Dewey argues, is flawed because it does not have the conceptual resources to draw the distinction between those episodes which can be described as genuine episodes of knowledge and those which cannot. This is particularly evident in the case of inquiry. The standard epistemological move is to abstract from the concrete conditions of inquiry – in particular, from its antecedent conditions – and, then, to frame the problem of knowledge as that of understanding "the relation of thought as such to reality as such" (MW2, 302). Clearly, some level of generalization is needed for a theory of knowledge to be possible. However, generalization is a powerful conceptual tool, and should be handled with care. Indeed, its results can easily be misleading if it is carried out without due regard to its possible distorting effects over its subject-matter.

The notion of internal conditions of inquiry is, thus, a reminder of the intrinsic limits of generalization. Its function is to call attention to the bounds of inquiry, i.e., to those conditions that make a course of reflection possible. Dewey is explicit on this point:

> Generalization of the nature of the reflective process certainly involves elimination of much of the specific material and contents of the thought-situations of daily life and of critical science. Quite compatible with this, however, is the notion that it seizes upon *certain* specific conditions and factors, and aims to bring them to clear consciousness – not to abolish them. While eliminating the particular material of particular practical and scientific pursuits, (1) it may strive to hit upon the common denominator in the various situations which are antecedent or primary to thought and which evoke it; (2) it may attempt to show how typical features in the specific antecedents of thought call out to diverse typical modes of thought-reaction; (3) it may attempt to state the nature of the specific consequences in which thought fulfills its career (MW2, 303).

Accordingly, a legitimate generalization is that which not simply preserves, but also highlights the intermediate character of inquiry. In doing so, since inquiry is, by definition, a derivative phase of experience (in the sense clarified above), such generalization does not erase its conditions of possibility and conceivability.

The same point can be made in a slightly different manner. It can be said that Dewey's theory of knowing has a stronger expressive power than traditional epistemological approaches in that, contrary to the latter, the former has the theoretical resources to offer a more fine-grained account of the relation between the elements of knowledge. Instead of thinking of them as "thought as such" and "reality as such", Dewey's account conceives of them as two poles of a reflective activity which originates from antecedent, problematic conditions and eventuates in a reconstructed, harmonic situation. Inquiry can be, therefore, analyzed in its specific sequence of development, and its features as a distinctive form of behavior can be clarified in light of their relation to its temporal character. In addition, by insisting on the structural differences between inquiry and its conditions, Dewey's account of knowing activity can also provide a justification of the metaphilosophical assumption that reflective and non-reflective activities cannot be described in the same language. The specificity of inquiry as a form of behavior requires a specific conceptual apparatus in which to describe it.

Before moving on to the next section, I would like to call attention to two relevant consequences that can be derived from the notion of internal conditions of inquiry. The first one has to do with the history of the reception of Dewey's philosophy. Traditionally, pragmatism has been charged with being exclusively concerned with the consequences of inquiry. That charge is usually associated with the idea that pragmatism promotes a form of raw utilitarianism, to the effect that the success of a certain course of action is not simply a reliable sign of its validity, but, more strongly, coincides with it.

In a more refined and theoretically interesting way, the same point has been made by Brandom who has repeatedly stressed that the main difference between his own inferentialism and classical pragmatism relies on the fact that, contrary to the latter, the former acknowledges the ineliminability of the semantic role played by what he calls "antecedents upstream" for the understanding of conceptual content. In his *Articulating Reasons*, this point is formulated with the following words:

> Pragmatists of the classical sort, by contrast, make the [...] mistake of identifying propositional contents exclusively with the *consequences* of endorsing a claim, looking downstream to the claim's role as a premise in practical reasoning and ignoring its proper antecedents upstream.[...] Yet one can know what follows from the claim that someone is responsible for a particular action, that an action is immoral or sinful, that a remark is true or in bad taste, without for that reason counting as understanding the claims involved, if one has no idea when it is appropriate to make those claims or apply those concepts. Being classified as AWOL does have the consequence that one is liable to be arrested, but the specific circumstances under which one acquires that liability are equally essential to the concept (Brandom 2000, 66).

I am not interested in discussing the technicalities of Brandom's inferentialism, neither am I willing to deny the differences between the two approaches – though my semantic reading of Dewey's notion of inquiry is strongly sympathetic to many of Brandom's views. What I am interested in is the general schema of Brandom's argument, and, more precisely, his statement that classical pragmatism downplays the import of the antecedent conditions.[23] In the light of what has been said, it should be clear that the consequence-focused interpretation of Dewey's logical theory that

[23] A word of clarification is in order here. I am aware that in the passage just quoted Brandom is dealing exclusively with the *semantic* import; so, properly speaking, his argument does not have any particular bearing on the epistemological questions under discussion here. Nonetheless, I believe that his remarks can be easily generalized. Obviously, Brandom cannot be held responsible for this kind of generalization.

Brandom advances is untenable. Not only Dewey emphasizes the importance of the antecedent conditions of inquiry, acknowledging the constraints that they put on the inquiry which stems from them; he also stresses their guiding role in directing reflection. Consequences are undoubtedly important in Dewey's account, but they are relevant only to the extent that, and insofar as, they are related to the whole of inquiry.

Dewey is extremely clear on this point. In *Reconstruction of Philosophy* he explicitly addresses some of the confusions that are traditionally imputed to pragmatism. Referring to the idea of truth as satisfaction, Dewey remarks that its theoretical significance has often been misunderstood by critics and interpreters. Indeed, that idea has usually been read as an attempt to redefine the notion in subjectivistic terms, so as to signify the subjective satisfaction for the conclusion of the process of inquiry. In reality, Dewey's conception of truth as satisfaction is not open to that criticism. Dewey formulates this point with the following words:

> the satisfaction in question means a satisfaction of the needs and conditions of the problem out of which the idea, the purpose and method of action, arises. It includes public and objective conditions. It is not to be manipulated by whim or personal idiosyncrasy. Again when truth is defined as utility, it is often thought to mean utility for some purely personal end, some profit upon which a particular individual has set his heart [...]. As matter of fact, truth as utility means service in making just that contribution to reorganization in experience that the idea or theory claims to be able to make (MW12, 170).

I will deal with Dewey's conception of truth in the final section of this chapter, so I do not dwell on this issue any further here. For now, it suffices to note that Dewey explicitly acknowledges that inquiry is subjected to friction, and that part of that friction comes from the rear, so to say, that is, from its antecedent conditions. The criticisms *à la* Brandom are, therefore, misplaced.

The other remark that I would like to make is intended to highlight an important aspect of Dewey's notion of inquiry that has usually gone unnoticed. As has been noted by some interpreters, Dewey's conception of inquiry seems to rule out all cases of unsuccessful inquiries. Recall his definition of inquiry: "[i]nquiry is the controlled or directed transformation of an indeterminate situation into one that is so determinate in its constituent distinctions and relations as to convert the elements of the original situation into a unified whole" (LW12, 108). Actually, such definition seems to imply that only those courses of reflective activity which succeed in bringing about a determinate situation should be called inquiry. So, it has been objected that Dewey's definition of inquiry is, at best, incomplete and, at worst, unsatisfactory (Browning 2002, 168).

Brown has argued that the easiest way out of this difficulty is to understand Dewey's notion of inquiry as an intrinsically normative notion – or, to use his own words, as "a success term" (Brown 2012, 276). Properly speaking, he remarks, "engaging in inquiry requires exercising proper control and the eventual settlement on a determinate situation". It follows, therefore, that any course of activity that is not directed at, and not succeeds in, transforming in a deliberate and controlled way an indeterminate into a determinate situation should not be considered an inquiry but, rather an attempt "to grope haphazardly" (Brown 2012, 276). In this sense, there is no unsuccessful inquiry: inquiry is, *properly speaking*, crowned with success.[24]

I agree with Brown on this point, but I think that something more can be said in favor of his reading, which he leaves quite undeveloped. As I have remarked, Dewey's theory of knowing is theoretically and expressively richer than traditional epistemology. I have also stressed that Dewey's approach is strongly anti-subjectivistic, to such an extent that it rejects the epistemological vocabulary centered around the notions of doubt, belief, knowledge, and so on. Finally, I have just insisted on the friction exerted on inquiry by its antecedent conditions. All this implies that inquiry is not to be understood as an activity that a subject can undertake *ad libitum*. Indeed, for an inquiry to be possible, its antecedent conditions must be thus made and arranged to provide a possible way to transform the original situation that they constitute into a unified whole. Consequently, any course of activity that takes the form of an inquiry is structurally successful, in the sense that its structural features are prone to bring about a successful reconstruction of the problematic situation.[25]

Stated in other words, while traditional epistemological approaches do not have the conceptual resources to adopt a strong evaluative stance on the various acts of knowing, Dewey's account of knowing activity is rich enough to provide such a

---

[24]This way of speaking is likely to seem awkward. Intuitively, we feel that it is safe to say that an inquiry may go wrong. In addition, it sounds strange to say that we do not know whether or not something can legitimately count as an inquiry until it comes to an end. In this spirit, Burke has suggested slightly modifying Dewey's definition of inquiry: inquiry should be defined not as "the controlled or directed transformation of an indeterminate situation *into* one that is so determinate [...]" – as Dewey puts it – but rather as the controlled or directed transformation of an indeterminate situation *toward* one that is so determinate [...]" (Burke 2009, 160). In doing so, he aims to make room for the possibility of inquiry to go wrong. In any case, as Browning has correctly put it, "the major problem for our [...] purposes is that of determining whether the 'definition' [of inquiry] is proposed as introducing a systemic use of the term *inquiry* or as appealing to a presystematic use that is sharable by others" (Browning 2002, 169). I will get back to this issue in Sect. 3.5.

[25]A remark can be useful here. It might be objected that an inquiry can turn out to be unsuccessful not because of some kind of structural shortcomings, but rather because of a lack of responsibility on the part of the inquirer. For instance, she may decide that she is not really interested in solving the problem that has called out her reflection. So, thus the argument goes, even though that particular inquiry is *structurally* or *potentially* successful since its antecedents conditions are thus made and arranged to show the inquirer how fix the problem, the inquiry does not succeed in reconstructing the original problematic situation. This argument has some plausibility only because it restricts the scope of the notion of internal conditions. First of all, it is trivially true that the inquirer is an element and component of the situation; this means that her lack of responsibility counts as an impeding and negative (structural) feature of the whole situation. Secondly, it is important to remind that the consequences of inquiry are elements that contribute to that normative evaluation. In this sense, as Brown has correctly pointed out, "to end the process of inquiry before its proper conclusion would be to give up on the inquiry before it is complete". And, in cases like this, I agree with Brown that "Dewey then would not say that the inquiry had terminated unsuccessfully, which implies a process came to an end, so much as merely ceased and dissolved prematurely" (Brown 2012, 276).

normative standpoint. As is quite well known, the notion that bears this normative load is that of situation. The next section is entirely devoted to the analysis of that concept, with the aim of shedding light on the idea of normativity of inquiry.

3.4 The Normativity of the Situation

The notion of situation is the hinge on which Dewey's account of inquiry turns. As repeatedly remarked, inquiry is defined as the transformation of an indeterminate situation into a determinate and harmonic one. But what is a situation? And, more relevantly, what is the relation between the concept of situation and that of inquiry?

We have seen above, in Chap. 1, that situation, like experience, is an infinity-word. This means that it purports to refer to the broad context of activity in which the objects with which that activity is concerned are framed, and from which they draw their significance. This insight – which is originally formulated in the Introduction to the *Essays in Experimental Logic* – is confirmed and restated in the *Logic*:

> What is designated by the word 'situation' is *not* a single object or event or set of objects and events. For we never experience for form judgments about objects and events in isolation, but only in connection with a contextual whole. This latter is what is called a 'situation' (LW12, 72).

Situation is, therefore, a context of activity. Since this notion is mainly used by Dewey in relation to inquiry, it can be said that situation refers to those contexts of activity that lead – or, better said, have the potentialities to lead – to processes of reflective and deliberate transformation of the environment.[26]

It is important to remark the highly complex 'modal' nature of the notion of situation. The fact that the indeterminate situation has certain, distinctive features is certainly necessary in order for an inquiry to be possible, but the mere existence of those features does not necessitate the agent to undertake a course of reflection. This is an aspect of Dewey's philosophy that has usually gone unnoticed, but which is pivotal for understanding his theory of inquiry. The distinction between indeterminate and problematic situation – which is quite elusive in the context of Dewey's logical thought – is intended precisely to make room for the idea that the act of undertaking a process of inquiry depends on the adoption of a reflective attitude, which is a decision that the inquirer is responsible to make. At the same time, however, the indeterminate situation is the basis or ground of the problematic situation, since they are two different yet essentially interrelated moments within a single process of inquiry which eventuates in a new determinate situation.

---

[26]In *Knowing and the Known*, the connection between situation and inquiry is explicitly stated. Situation is here defined as the "[e]vent as subjectmatter of inquiry, always transactionally viewed as the full subject-matter; never to be taken as detachable 'environment' over against object" (LW16, 71).

What about, therefore, the relation that holds between the (three) moments through which the process of inquiry is articulated? As a biologically-grounded process, inquiry is a temporal activity. As noticed above, this does not simply mean that inquiry takes time to be completed, which would be a trivial point; it also means that its subject-matter changes throughout its developmental stages. Indeterminate, problematic and reconstructed situations are, thus, to be understood as the three distinct ways in which the subject-matter presents itself during the course of inquiry. And yet, those three phases are not separate or independent one from the others: there is a unity that runs through them, which makes them parts of a broader totality. So, the issue of understanding the relation between inquiry and situation becomes that of understanding the nature of the underlying unity that makes it possible for an activity to articulate an indeterminate subject matter into a problematic one, and then to transform the latter into a reconstructed, determinate and harmonic situation.[27]

Let's start, therefore, from the beginning – that is, from the analysis of the distinctive features of the indeterminate situation. Here is how a plausible definition may look like: a situation is indeterminate when the *significances* of the objects with which the agent in her life-behavior is concerned are obscure, confused or conflicting; which means that the habits of the agent – those habits on which her behavior relies – are not effective in conducting the course of activity in which she is currently engaged to a successful conclusion. Stated in other terms, in an indeterminate situation some of the concepts that the agent uses in her transactions with the environment are not attuned to their objects: as a consequence of that, objects lack clear significance, while concepts do not have fixed meaning.

This point can be illustrated by means of a very simple example, whose philosophical import can be easily generalized to the case of highly refined scientific inquiries. Take, for instance, a situation in which you need to break a nut and you do not have a nutcracker at hand. You see an ashtray on the table, and you ponder whether it can be used as a nutcracker. In this situation, the significance of the ashtray is partially undetermined in light of the purpose at stake. Clearly, the greatest part of its significance is fixed since you do no doubt about its being an object that persists in time; neither you doubt about its being made of, say, copper. However, you do not know whether it is hard enough to break a nut: so, while you are ready to attribute to it the properties of being a material object and of being made of copper, you are not ready to attribute to it the property of hardness. In this sense, the

[27]Browning has argued that inquiry should be viewed as "consisting of a changing succession or stream of situations" (Browning 2002, 171). Browning's account has been critically analyzed by Burke. Though he acknowledges that there are passages in Dewey that support Browning's reading, he remarks that "the fact that the character of the situation changes does not mean that the numerical identity of that situation changes". And then he notices: "note the need on Browning's part (2002, 171) to introduce some kind of 'unique and relevance-guided quality' pervading the alleged succession of situations in order to thread together what has just been sliced into parts. There will be no need to postulate this extra 'mysterious tie' if the required unique and relevance-guided quality is provided already by the one situation undergoing transformation" (Burke 2009, 164). The unity of the situation is internally composed of a variety of different phases. I wholly agree with Burke on this point.

significance of the object and the meaning of the corresponding concept have to be established.[28]

As is evident, Dewey's conception of indeterminacy is at odds with standard contemporary philosophical approaches, and for a twofold reason. First of all, we are used to think of the quality of indeterminacy – along with other notions such as that of vagueness, for instance – as an epistemic notion. According to the standard view, the world is not vague; what is vague are, rather, our thoughts on the world. There are no vague properties; there are only vague predicates.

This is not the case for Dewey, however. It is quite clear from what he writes in the *Logic* that he is not against the idea of vague properties. Nonetheless, he does not need to make use of the idea of vague properties in order to ground his view of the objective nature of indeterminacy; a weaker version is sufficient for that purpose. As he remarks, in an indeterminate situation "[e]ven were existential conditions unqualifiedly determinate in and of themselves, they are indeterminate in *significance*," that is, "in what they import and portend in their interaction with the organism" (LW12, 110). Now, such indeterminacy is not epistemic, strictly

[28]Dewey's non-epistemic account of the indeterminate situation seems to entail the conclusion that a property, say, the property of hardness, happens to be embodied in a particular object only after that object is actually put to test – in this case, it is used to break the nut. That line of thought is reminiscent of Peirce's first formulation of the pragmatic maxim – in *How to Make Our Ideas Clear* (1878) – and is extremely problematic in that it attempts a reduction of a bunch of interrelated normative notions (laws, concepts, success) to non-modal concepts such as concrete empirical verifiability. In other words, in order to avoid an epistemic account of situation, Dewey's position seems to collapse into an extreme and untenable form of nominalism. In this regard, two points have to be noted. First of all, the distinction between environment and world can be used here to clarify the view that Dewey is struggling to outline. Dewey would be happy to concede that, at the level of existence, the object possesses a certain property independently of the process of inquiry through which the significance of the thing comes to be fixed (LW10, 110). This is rather obvious from a Deweyan perspective, since issues of existence are, by definition, different from issues of meaning (or significance). In so doing, the realistic insight that the possession of a property is an objective affair is preserved within Dewey's conceptual framework, thus blocking any possibility to jump to idealistic conclusions. Secondly, the attribution of a property to an object is an example which supports strongly realistic insights, and, consequently, it rather naturally leads to a representationalist account of cognition. Accordingly, it is a bad example for the purpose of clarifying Dewey's position. As said above, it would be absurd to call into question the fact that an object has the properties that it has, independently of the process of inquiry through and by which the inquirer comes to discover them. In cases of inquiry such as that of the nutcracker, the goal of inquiry is the acquisition of knowledge about some state of affairs. Hence the intrinsic plausibility of the epistemological approach: since things are completely determined in themselves, and since the goal of an act of inquiry is that of grasping the content of objective reality, inquiry is adequate or successful to the extent that it represents in a faithful way the state of affairs that it purports to depict. However, there are other examples in whose light that realistic insight loses its grip. Think, for instance, at all those cases in which inquiry requires an active manipulation of the elements of the situation for the purpose of changing the existential conditions that called out inquiry. In those cases, the goal of inquiry is not that of representing something which we take to be already out there, but rather that of bringing about new existential conditions. Here the realistic language of representation is misplaced.

speaking. It is an objective quality of the kind of responses that the agent is ready and able to give to the stimuli of the environment.

Dewey is trying to find a difficult middle way between the Schylla of metaphysical realism and the Charybdis of epistemological subjectivism. Because of his externalism, Dewey rejects the idea that the indeterminacy of the objects of the situation has to be understood "in a 'subjective' sense" (LW12, 110). Accordingly, the situation is not *epistemically* undetermined: "[i]t is the *situation* that has these traits," Dewey remarks, and not simply our beliefs about it. Even more clearly, he states that "*[w]e* are doubtful because the situation is inherently doubtful" (LW12, 109). It is the doubtful, objective quality of the situation that makes it possible for the agent to be in a state of doubt. This is a semantic argument: it is the situation that fixes the semantic content of our beliefs. If a state of doubt were completely disconnected from the existential features of the situation, it would count as a psychological state of distress rather than as an epistemic state of withdrawal from belief, and its contents would not be attuned to the environmental conditions to which they purport to refer.

At the same time, however, Dewey also rejects the idea that the indeterminacy of the objects of a situation has to be understood in a strong metaphysical fashion. By definition, inquiry is centered on the particular inquirer who carries it out: indeterminacy is a quality of a specific, individual transaction between the inquirer and her environment; it is not a quality of the objects taken in themselves. The indeterminacy of the objects of a situation is, therefore, an agent-dependent property. It is the property that some objects have for an agent who is engaged in a particular course of activity at a certain time. Any attempt to extend that property beyond the boundaries of the situation in which it takes place is illegitimate.

The latter remarks lead directly to the other aspect in which Dewey's conception of indeterminacy goes against the grain of much of contemporary philosophy. Once again, that disagreement is a consequence of Dewey's metaphilosophical assumption that the category of practice has a theoretical and explanatory primacy over that of object. Though it is correct to say that objects are indeterminate – Dewey himself states that the existential conditions are indeterminate in significance (LW12, 110) – properly speaking it is the whole situation which is indeterminate, while the objects are indeterminate only indirectly, as objects belonging to that particular situation. This point can be easily grasped if framed in practical terms. When seen from that perspective, it is clear that the indeterminacy of the objects originates from the fact that they do not hang together *within* the context of a particular life activity. Now, "hanging together" is a property of the whole situation, not of its elements.

Intuitively, there is room to resist such conclusion. One possible way to counteract Dewey's argument – thus defending the primacy of the category of object over that of practice – would be to understand the indeterminacy of the situation as a higher-order property of objects. However, Dewey explicitly rejects this view. He states that the quality of a situation – just like the quality of a painting – "is not a property which it possesses in addition to its other properties"; rather, it is something which "externally demarcates" that particular situation from the others, and which" internally pervades, colors, tones, and weights every detail and every relation" of the context of activity with which the agent is concerned (LW5, 245).

Borrowing the expression from Santayana, Dewey calls the qualities of the situation 'tertiary qualities'. Now, tertiary qualities are not reducible to primary and secondary qualities. The easiest way to characterize the difference between them is in terms of their bearers: while the bearers of primary and secondary properties are objects, the bearers of tertiary qualities are situations.

This consideration might seem trivial, but it is not. When the categorial difference that exists between situation and object is paid attention to, it is clear that the way in which a tertiary property belongs to a situation cannot be identical to that in which a primary or secondary property belongs to an object. Dewey is explicit in this regard: "[t]he phrase 'tertiary quality' [...] does not refer to a third quality like in kind to the 'primary' and 'secondary' qualities of Locke and merely happening to differ in content" (LW12, 75). The difference lies, rather, in the very structure of the relation that holds between the bearer and its property.

I would like to call attention to two features of that relation which are particularly relevant to the understanding of Dewey's philosophy of inquiry. First of all, tertiary qualities are holistic properties, that is, properties that are spread all over the objects of the situation that is their proper bearer. Dewey remarks that "a tertiary quality qualifies *all* the constituents to which it applies in thoroughgoing fashion" (LW12, 75). So, while properties like hard, red, sweet, are associated with some aspect of a specific object within an experience – thus enabling the inquirer to discriminate between them and, consequently, to use them in a proper way – tertiary properties work differently.[29] Indeed, it is possible to compare a primary or secondary quality with another quality of the same kind, but the same operation is impossible in the case of tertiary qualities. By qualifying all the elements of a situation, they provide a condition of identification of certain objects as the objects of *that* particular situation. Here is how Dewey formulates this insight:

> By the term situation in this connection is signified the fact that the subject-matter ultimately referred to in existential propositions is a complex existence that is held together, in spite of its internal complexity, by the fact that it is dominated and characterized throughout by

---

[29]The idea that Dewey wants to convey can be clarified with a simple example. Imagine that you are presented with a series of pictures. You are sitting at your desk, and a person puts in front of you one picture after the other. Suppose that all these pictures portray a different aspect of a single, ordinary object. The pictures are presented in a fixed order, from left to right, and from top to bottom; moreover, the previous picture is not removed from the desk when the next one is presented. Now, it is very likely that, at a certain moment, you will recognize the object portrayed in the pictures. From that moment on, something changes: what was previously experienced as an obscure series of pictures acquires significance; in Dewey's own words, things now hang together (LW12, 109). As a consequence of such transformation, the situation changes in its overall quality. It stops being an indeterminate situation in which you cannot understand how things are related one to the others, and become a harmonic whole in which any element is intrinsically connected to those that precede it, as well as to those that will follow or stem from it. Things are now experienced as being in their proper place, so to say. And yet nothing more is added to the series than the single picture which, so to say, 'unlocks the significance' of the whole situation.

> a single quality. By "object" is meant some element in the complex whole that is defined in abstraction from the whole of which it is a distinction. The special point made is that the selective determination and relation of objects in thought is controlled by reference to a situation – to that which is constituted by a pervasive and internally integrating quality, so that failure to acknowledge the situation leaves, in the end, the logical force of objects and their relations inexplicable (LW5, 246).

Secondly, being holistic, tertiary qualities are also exclusive. Up to now I have spoken as if it were possible for a situation to have many different qualities. That was a loose way of speaking. Assuming that different tertiary qualities may inhere to one situation amounts to a categorial mistake due to a lack of comprehension of the radical difference existing between objects and situations. An object has different primary and secondary qualities since, if it is not an abstract entity, it is in many different ways; on the contrary, a situation is characterized by its distinctive quality. Leaving aside the issue whether this insight can be generalized to any kind of property whatsoever, it is clear that in the case of tertiary qualities, substance and property are essentially correlative notions: a tertiary quality is the way the situation is, while the situation is what bears that quality, and nothing else. Consequently, the situation *is* its distinctive quality. Tertiary qualities provide, therefore, the criteria of identity of situations, as well as their criteria of individuation. In order to identify a specific situation, one has to turn to its distinctive quality – full stop.

This assumption is theoretically powerful. For instance, it implies that no matter how detailed a description of the objects of a situation might be, it cannot provide a satisfactory identification of that situation. The identity of the situation is immediately experienced rather than grasped through an act of analysis, since any act of analysis presupposes a situation in which to take place. The notion of tertiary *quality* aims precisely to highlight the fact that the unity of a situation is a property had or felt by the agent who is engaged in that particular course of activity. I will come back to this point in a moment.

In addition, the idea that tertiary qualities provide the criteria of identity of situations enables Dewey to justify his radical pluralism about inquiries. Take, for instance, the following passage: "[t]he pervasively qualitative", Dewey remarks, "is not only that which binds all constituents into a whole but it is also unique; it constitutes in each situation an *individual* situation, indivisible and unduplicable" (LW12, 74). Here it is clear that Dewey considers tertiary qualities non only as the criteria of identity and identification of situations, but also as the ground of their individuality. That this is the case is confirmed by his statement that "a situation is a whole in virtue of its immediately pervasive quality" (LW12, 73). Now, since tertiary qualities are unique, so are the situations that are their bearers.[30]

---

[30]The same insight can be formulated by acknowledging that any situation is centered on a specific, individual inquirer who is engaged in a specific course of action, determined by her biological and cultural endowment – i.e. the meanings and significances to which she responds. From such an assumption a genuine plurality of individual and independent situations follows. Indeed, a situation is, by definition, a-situation-for-someone. This means that the very same objective conditions – say, a door close without the key in keyhole – may well have different significances for an agent A, who

In order to account for his pluralism about inquiries, in his *Logic* and in the essay *Qualitative Thought* Dewey introduces a new conceptual distinction between universe of discourse and universe of experience. Such distinction is substantially homologous to the one between infinity (or zero) words and proper terms. Indeed, just as the use of proper terms is possible only within the context of an experience or a situation, similarly the universe of discourse necessarily depends on the universe of experience. However, the new distinction is theoretically useful in that it helps bring to the fore the normative role played by the situation with regard to the determination of its objects. "A universe of experience", Dewey remarks, "is the precondition of a universe of discourse" since "[w]ithout its controlling presence, there is no way to determine the relevancy, weight or coherence of any designated distinction or relation" (LW12, 74). It is the felt quality constituting the situation that gives sense to the different operations – symbolical as well as physical – through which the objects are handled. As Dewey notices, "[d]iscourse that is not controlled by reference to a situation is not a discourse, but a meaningless jumble" (LW12, 74). It is only by paying attention to the broader context of practice in which the agent is engaged that her action can be normatively attuned to the end-in-view that defines that practice.

An example may be useful. Consider, for instance, a math problem that you are asked to solve. You start reflecting on it, you write down some equations, then you work on them, and so on. All your operations seem to perfectly hang together: any operation is done for the purpose of preparing the way for the next one, and, taken together, they are expected to lead to the solution of the problem. At a certain moment, however, you lose the thread of the discourse: all of a sudden, the operations that run smoothly one into the other become disconnected. The quality of the situation has changed: the normative relations that held its different objects and elements together are now broken, and without their controlling presence the inquirer lacks the normative guidance that is necessary to accomplish her goal.[31]

Tertiary qualities perform, therefore, three different yet interrelated functions: (1) they provide the conditions of identity of situations; (2) they provide their conditions of individuality; (3) they provide the normative bonds that, by keeping

knows the trick to unlock the door, and for an agent B, who is actually trapped in the room, trying to get out. This conclusion is, once again, in agreement with Dewey's idea of the theoretical and explanatory primacy of the category of activity over that of object: it is only within a situation that objects acquire their distinctive significance – their significance in light of the end-in-view which the agent aims to achieve. Consequently, the mere fact of the existence of the same environmental conditions does not entail that two agents facing them are engaged in the same activity. Rather the contrary, the situations with which they deal – even though it can be said that they share the same subject-matter – are essentially distinct and different.

[31] In *Qualitative Thought* Dewey resorts to the popular, unrefined notion of intuition to formulate this insight. He writes: "in its popular, as distinct from refined philosophic, usage [intuition] is closely connected with the single qualitativeness underlying all the details of explicit reasoning. Reflection and rational elaboration spring from and make explicit a prior intuition [...]. Intuition, in short, signifies the realization of a pervasive quality such that it regulates the determination of relevant distinctions" (LW5, 249).

the various elements of a situation together, constitute an internally articulated whole that is capable of undergoing changes through time, without losing its identity and individuality.[32]

The last remark is of the utmost importance. As hinted at above, the kind of normativity that is provided by tertiary qualities is felt rather than grasped as an object of discourse (LW12, 73–74; see also LW5, 248). Dewey is explicit in stressing that, though the pervasive quality of the situation "surrounds and regulates the universe of discourse", the former "never appears as such" within the latter. The specific quality that constitutes a situation cannot, by principle, be *stated* in a discourse which has as its subject-matter the objects of that situation, and as end-in-view its reconstruction. A tertiary quality exerts its normative function from the outside, so to say: the inquirer "feels" – but Dewey warns not to read this notion in a psychological manner – that her operations are internally related, and are all appropriate as means to the end.

I think that there are two reasons why Dewey insists on the fact that the normativity of a situation is to be of the nature of a quality. The first reason has to do with the criteria of identity of a situation. As is well known, it is notoriously difficult to provide criteria of identity of objects over time. At the same time, however, it seems evident that any activity goes through different phases while retaining its identity and individuality. Now, Dewey argues that the identity of the activity through its phases is guaranteed by their being felt as internally related moments of a broader totality.

The very same insight lies at the basis of Dewey's notion of *an* experience, as formulated in the third chapter of *Art as Experience*. In this work, Dewey stresses the essential connection between individuality and felt quality, and in doing so he also emphasizes the 'aesthetic' – i.e. immediately and directly grasped – nature of the principle of identity of an experience. Here is what Dewey writes in this regard. Firstly, he remarks that "[a]n experience has a unity that gives it its name, *that* meal, that storm, that rupture of friendship"; then he adds that "[t]he existence of this unity is constituted by a single *quality* that pervades the entire experience in spite of the variation of its constituent parts" (LW10, 44); finally, he concludes that "*an* experience of thinking has its own esthetic quality" (LW10, 45). The distinctive feature of *an* experience is that "every successive part flows freely, without seam and without unfilled blanks, into what ensues": as Dewey puts it, referring explicitly to the underlying unity of reflective activities, it is correct to say that thinking goes on "in trains of ideas", provided that the latter are understood as "phases, emotionally and practically distinguished, of a developing underlying quality" (LW10, 44). And

[32]In Pappas (2014, 2016) the author distinguishes between nine different functions of the qualitative in inquiry. I focus on three. The difference between the two lists is only superficial, however. Pappas's distinctions are more fine-grained than mine – for instance, he clearly distinguishes between "The Qualitative as the Background that Unifies and Demarcates the Situation in which Thinking Occurs" and "The Qualitative as the Background that Gives Continuity to Thinking" (Pappas 2016, 446 and 449) – but there is no real disagreement between the two accounts about the general functions performed by tertiary qualities in inquiry.

it is the very fact of being part of this qualitative unity that gives to its different phases their proper character and significance: "[a]s one part leads into another and as one part carries on what went before, each gains distinctness in itself" (LW10, 43).

I suggest, therefore, that Dewey's solution to the issue concerning the criteria of identity of a situation is to emphasize the difference between the category of situation and that of object, and to conclude that objects and situations have different criteria of identity. Indeed, while it is possible to formulate the criteria of identity of an object, as well as to raise a challenge over its identity, this is not feasible in the case of situation. Every tertiary quality is "unique and inexpressible in words", and cannot be made explicit in a discourse. Consequently, there is no room for raising the problem of its identity: that is beyond our possibility of meaningful assertibility.

The second reason has to do with how to counteract the rule-following problem. It should be clear that, thanks to the idea of tertiary quality, the regress of justification can easily be blocked. However, in Dewey's practice-oriented framework the regress of justification is not blocked by discovering a rule that can be followed without relying on another, more fundamental rule; rather, it is blocked by acknowledging that the agent feels that she is doing the right move at the right time and for the right purpose. The agent feels that her action are normatively adequate, and she does not need any justification apart from that feeling of adequacy. She has reached the bedrock, and her spade is turned.

I do not agree, therefore, with Burke when he argues that Dewey's concept of tertiary quality plays a foundational role in Dewey's epistemology. In his *What is a Situation?* Burke reads the notion of tertiary quality in continuity with the Cartesian project of finding some element that is "immune to the appearance/reality distinction". Now, he is right in pointing out that a quality – since it is felt – cannot be different from how it appears. He is also right that the normative guidance of the felt quality supplies the inquirer with a reliable indication of what response is adequate in any given instance. Nonetheless, the conclusion he draws from those premises – namely, that tertiary quality provides an *epistemological* foundation to our knowledge – is, in my opinion, wrong. For two reasons. The first one has to do with Burke's understanding of the notion of situation. He writes that "Dewey used the term 'situation' to refer to this immediate pervasive quality of an experience" (Burke 2000, 109). I reject this reading, on a textual basis. It is true that Dewey defines situation as the "contextual whole" in which we experience objects and events. Nonetheless, it seems clear that situation should not be conceived as a feature of experience, but rather as a different word (an infinity word) by which he refers to the same transactional whole to which he refers by 'experience'.

The other reason is philosophical, and is more relevant for the purposes of this present discussion. Burke advances an epistemological reading of the notion of situation. He says that the qualitative immediacy of the situation "serves the same role in a given inquiry that the cogito was suppose to play in a Cartesian epistemology". And then he adds:

> In any given instance, this qualitative immediacy of a possibly complex situation is a solid epistemic foundation in that it is as immediate and obvious and indubitable as is Descartes' s own thinking is to himself in the particular inquiry described in *Meditations on First Philosophy* (Burke 2000, 112).

In my reading, on the contrary, the notion of tertiary quality plays a semantic rather than an epistemological function. Tertiary quality is not the first premise of a reasoning, as Burke seems to imply by suggesting a strong analogy between Descartes's cogito and Dewey's tertiary quality. Rather, it is the comprehension of what is going on in the course of inquiry, which provides the normative guidance for future operations. In this light, it might be conceded to Burke that tertiary qualities bear foundational weight, as long as foundational is not understood in epistemological terms.

We are now in the position to answer the question raised at the beginning of this section, namely, how it is possible that inquiry goes through different phases of inquiry – indeterminate situation, problematic situation, harmonic situation – without resulting in a series of disconnected and unrelated members. The answer is that any inquiry is characterized by a distinctive tertiary quality, and because of such underlying tertiary quality running through all the different objects of the corresponding situation, as well as through the different activities undertaken by the inquirer, reflective behavior is made possible.

Inquiry is, therefore, the process through which the original, indeterminate situation is firstly articulated and stated in a problematic situation, and then reconstructed into a harmonic situation.[33] It is the same, unique situation that undergoes a process of internal modification. Dewey is rather clear on this point: it is the same subject-matter that is gradually transformed through inquiry, thus taking different forms and exerting different functions. In *Qualitative Thought*, Dewey states that inquiry makes explicit what is contained in the original pervasive quality. The possibility of such continuity of subject-matter is guaranteed by the persistence over time of the tertiary quality which defines and constitutes the specificity of that particular, individual inquiry. In light of all this, inquiry is to be understood as a series of internally connected, distinct temporal phases held together by a quality that provides normative guidance. This kind of aesthetic, felt normativity accounts for the relations that exist between the three phases of which any inquiry is composed.

In the next section, I will turn attention to the closing phase of inquiry, as well as to the notions – truth, warranted assertibility – that Dewey uses to characterize its distinctive features. In so doing, I will complete the analysis of the conceptual tools – both philosophical and metaphilosophical – that Dewey considers necessary to account for the temporal sequence of reflective activities.

[33] On the notion of articulation, applied to moral reasoning see Frega (2012).

3.5 Truth and Warranted-Assertibility

According to Dewey's definition of inquiry, the conclusion of the reflective process is a situation "determinate in its constituent distinctions and relations", to such an extent that it takes the form of "a unified whole" (LW12, 108). Contrary to the indeterminate and the problematic situation, the reconstructed situation does not present conflictual features: it is not divided within itself anymore. Obviously, its being a unified whole does not entail a lack of internal articulation. The notion of individuality – which Dewey uses to characterize the reconstructed situation – is distinguished from that of particularity or singularity (LW12, 126–127). While the latter refers to some aspect within a complex situation, the former refers to the unity of the whole situation. It is also evident that such unity is qualitative: it means that all the elements of the situation are pervaded by the same distinctive quality, as a consequence of which (a) they are all perceived as elements of that situation, and (b) they are attuned one to the others.

The end of inquiry is, therefore, to be understood as that phase of experience in which, being attuned one to the others, the reconstructed significances of the objects are capable of leading to overt action. When inquiry is completed, there is no need for further reflection, and, consequently, no need to continue to abstain from acting: the objects function as stimuli to action, and their significance is defined in terms of the meanings of the theory through which the problematic situation is reconstructed. More on that in the next chapter.

Dewey's account of the conclusive phase of inquiry has been widely discussed and criticized. The idea that pragmatism in general – and Dewey, in particular – ends up identifying the end of inquiry with the satisfaction of whatever desires an agent may have has been quite popular in the past, and has been often associated with the practical character of American culture. Leaving aside the trivialization of Dewey's views, it can hardly be denied, however, that Dewey is rather vague and unclear on this issue. So, though it is highly questionable whether Dewey means to substitute the *notion* of truth with that of warranted assertibility – I will get back to it momentarily – many readers and interpreters have read him as holding this view. Russell is one of them; Putnam's conception of truth as warranted assertibility under ideal conditions is another instance of a reading of this kind. The standard view is that Dewey purports to make truth an epistemological notion: by replacing truth with warranted assertibility, Dewey is credited as excluding (as meaningless) all those uses of "truth" which express the possibility that any assertion, no matter how well grounded and justified, could nonetheless turn out to be false.

In recent times, however, the standard view has been challenged, and attention has been called to the positive characterization that Dewey gives of truth (Capps 2018). It has been acknowledged, for instance, that Dewey accepts as the best *definition* of truth – at least from the logical standpoint – the Peircean view according to which truth is the opinion which is fated to be ultimately agreed by all who investigate; consequently, it is difficult to maintain that Dewey is a deflationist about truth, though it is beyond doubt that he is not very much interested in clarifying that notion (Capps 2017, 151). Similarly, since he admits some role for truth in logical theory, it seems implausible to me that he would be willing to *completely* replace the notion of truth with that of warranted assertibility. True, Dewey stresses that Peirce's definition of truth is "a definition of truth *as an abstract idea*"; and yet, I think that even abstract ideas – with whose definitions Dewey is not much concerned (Capps 2017, 151) – can play some sort of clarificatory function within Dewey's theoretical framework.

Capps has recently argued that Dewey's aim is neither to get rid of the notion of truth nor to provide a formal definition of it. Rather, his aim is to operationalize the notion, "linking it to the processes and practices of inquiry".[34] More precisely, according to this reading, Dewey's goal would be that of examining "how truth functions in actual inquiries that aim at resolving what he calls 'problematic' or 'indeterminate' situation" (Capps 2018, 45).[35] I completely agree with this way of framing the issue, which relies on the idea of a theoretical and explanatory primacy of activity – an assumption that I clearly share with Capps. In the following pages, therefore, I will elaborate on this idea, for the purpose of clarifying the main features of the conclusive phase of inquiry.

In order to reach this goal, I will first introduce Dewey's distinction between judgment and proposition, and I will insist on the difference between the instrumental nature of the latter and the definitive character of the former. Then, I will argue that such form of *unrevisability* is not necessarily problematic: it can seem problematic only in the case of scientific inquiries. Finally, I will advance two possible solutions to this problem.

On the contrary, I will not address the issue whether or not Dewey has a theory of truth; neither will I attempt to take a stance on the relation between truth and

---

[34]A similar view is formulated by Levine in his recent *Pragmatism, Objectivity and Experience.* Levine also highlights a structural similarity between truth and happiness. He writes: "for Dewey this logical conception does very little work in his thought. Dewey worried that focusing on it would divert our attention away from the methods by which our various inquiries actually fix belief and tempt us into reinstating a realist view of truth. But I think he had another worry. In his moral philosophy, Dewey argued that happiness is not "directly an end of desire and effort, in the sense of an end-in-view purposively sought for, but is rather an end-product, a necessary accompaniment, of the character which is interested in objects that are enduring and intrinsically related to an outgoing and expansive nature" (LW7, 198). To make happiness one's direct end is the surest way to not achieve it, for then one does not cultivate a genuine and direct interest in the kinds of objects that will, in fact, make one happy. I think he has the same thought about truth: instead of focusing on truth itself, we should – in light of our cultivated interests and habits – directly plunge into the objects of our concern. It is this that will produce truth, but as a by-product of, or accompaniment to, an inquiry that looks into objects in the right way" (Levine 2019).

[35]Capps has distinguished among three different interpretations of Dewey's theory of truth. The first one assumes that Dewey's goal is to get rid of the notion of truth, and to replace it with the concept of warranted assertibility. The second interpretation holds that Dewey is interested in rejecting the *words* 'true' and 'truth', but that he does not want to get rid of the corresponding notion. The third interpretation – which is the one Capps favors – argues that Dewey is dissatisfied with the idea of a *theory* of truth, and that he aims to provide a pragmatic elucidation of the concept of truth (Capps 2018, 41–43).

warranted assertibility. Suffice to say on this latter topic that, properly speaking, Dewey does not hold that warranted assertibility can be taken as a plausible substitute for truth: rather, he suggests that *knowledge* should be replaced with warranted assertibility. Here is what Dewey writes in this regard:

> If inquiry begins in doubt, it terminates in the institution of conditions which remove need for doubt. The latter state of affairs may be designated by the words belief and knowledge. [...] I prefer the words "warranted assertibility". [...] the term "warranted assertion" is preferred to the terms belief and knowledge [since it] is free from the ambiguity of these latter terms, and it involves reference to inquiry as that which warrants assertion (LW12, 15–16).

Now, it is indisputable that some connection exists between knowledge and truth; and yet it is also evident that they can be by no means identified – especially in the absence of a strong theoretical proposal that makes that identification at least plausible. Consequently, it is rather surprising to me that critics and scholars have read Dewey as advancing a thesis that, at least as far as I have been able to ascertain, is not supported by textual evidence.[36] I do not want to push this point too far, but it might be that the whole issue is in need of reconsideration.[37]

All this said, let's start the analysis with some preliminary, terminological considerations. As is well known, Dewey neatly distinguishes between proposition and judgment. This distinction is extremely idiosyncratic, being grounded in Dewey's idea of the intrinsically and constitutively temporal structure of inquiry. Within this framework of analysis, propositions are defined as those linguistic expressions that are employed in the course of reflection – they are the means through which inquiry

[36]It is likely that the confusion is due to Russell, who in *Meaning and Truth* writes that "[t]he theory which substitutes 'warranted assertibility' for 'truth' [...] is advocated by Dr. Dewey and his school" (Russell 1940, 289). See also Russell's reconstruction of Dewey's notion of truth in his contribution to the Schilpp volume on Dewey (Russell 1939). It is worth noting that in his more articulated reply to Russell's criticisms, the 1941 article *Propositions, Warranted Assertibility, and Truth*, Dewey explicitly remarks that warranted assertibility is introduced to take the place of the *word* 'knowledge', not of truth: "Mr. Russell refers to my theory as one which "substitutes 'warranted assertibility' for truth". Under certain conditions, I should have no cause to object to this reference. But the conditions are absent; and it is possible that this view of 'substitution' as distinct from and even opposed to *definition*, plays an important role in generating what I take to be misconceptions of my theory in some important specific matters. Hence, I begin by saying that my analysis of 'warranted assertibility' is offered as a *definition* of the nature of knowledge in the honorific sense according to which only true beliefs are knowledge. The place at which there is pertinency in the idea of 'substitution' has to do with *words*" (LW14, 168–169).

[37]For the sake of fairness, it has to be acknowledged that other interpreters have recognized that Dewey does not attempt to identify truth with warranted assertibility. So, for instance, Fesmire states that "Dewey aimed to incorporate the function of truth within the concept of warranted assertibility" (Fesmire 2014, 99). Similarly, Aikin and Talisse hold that, since from a pragmatic perspective there is no real difference between pursuing warranted beliefs and pursuing true ones, it follows that "*the true* is replaceable with *the warranted*". It is not easy to assess the validity of Aikin and Talisse's proposal, since it seems to imply that Dewey's account of knowledge can be traced back to the traditional analysis of knowledge as justified true belief. In any case, it is evident that Aikin and Talisse do not argue for the simple and direct identification of truth with warranted assertibility.

is carried out. Consequently, Dewey's notion of proposition has nothing to do with the contemporary conception of proposition as the cognitive content of a thought or a sentence. Judgment, on the contrary, corresponds to the conclusive phase of inquiry; or, better said, it *is* the conclusion of inquiry (LW12, 124).

I will discuss in detail the relations between propositions and judgment in the next chapter, so I am not going to tackle this issue here. What is worth remarking in the present context is what Dewey says about the truth-value of these two kinds of linguistic expression. Consider propositions first. In the section of the *Logic* dedicated to the analysis of propositions, Dewey writes that "propositions are to be differentiated and identified on the ground of the function of their contents as *means*"; and then he adds that "since means as such are neither true nor false, truth-falsity is *not* a property of propositions" (LW12, 287). Consequently, there is no doubt that Dewey adopts the view that propositions are not truth-bearers.

What about judgment? To my knowledge, in no place in *Logic* Dewey explicitly states that judgment is a truth-bearer – i.e., something which can be said to be true or false.[38] In any case, this position can be easily inferred by adding to what has been just said about propositions the following remarks. After noticing that judgment is the settled outcome of inquiry, Dewey stresses that judgment is to be distinguished from propositions; indeed, while the content of propositions is "intermediate and representative and is carried by symbols", judgment "has *direct* existential import" (LW12, 123). Now, since propositions are neither true nor false, and since judgment is said to be different from propositions, it seems safe to conclude that judgment is a genuine truth-bearer. In this sense, I take judgment and warranted *assertion* as synonyms.

These remarks are helpful in shedding light on how different Dewey's approach is from contemporary analytic accounts. In his framework of analysis, the difference between proposition and judgment is given *exclusively* in terms of the different place that they occupy within the course of inquiry: a linguistic expression is a judgment when it is uttered or formulated at the end of inquiry; the very same linguistic expression counts as a proposition if it is uttered or formulated during the course of inquiry. Clearly, such difference of place reflects a difference in function: judgment closes inquiry and leads to overt action; propositions provide procedural and material means to carry out inquiry. As Dewey writes in a letter to Bentley, "I used judgment, not to stand for a mental event, but for a factual end or termination definitely reached, and proposition for a process as still in process, and hence still tentative and relatively indeterminate". And, he adds, the distinction-connection between proposition and judgment is to be understood as "a specified case of the means-end distinction-connection" (1944.04.08, J. Dewey to A. Bentley).

[38] A remark here is needed. It is worth noting that if we accept the view that inquiry is a success term, judgment cannot be, strictly speaking, false. Indeed, according to this reading, the term 'inquiry' is meant to refer to successful inquiries only, and judgment is "the settled outcome of inquiry" which "is concerned with the concluding objects that emerge from inquiry in their status of being conclusive" (LW12, 123).

From our contemporary perspective, I think it would be quite natural to argue that the element that, added to a proposition, constitutes a judgment is justification. A plausible reconstruction of Dewey's position would thus run as follows: the propositions that we use during inquiry have a truth-value independently of our capacity to recognize it; at the end of inquiry, on the contrary, the linguistic expression is not simply true, but also warranted or justified. In other words, justification – which is the property that linguistic expressions acquire through the course of inquiry – enables us to grasp the truth of the proposition in the obvious sense that if a proposition is justified for a person, that person is likely to hold it true. This is close to the idea that true belief is not enough to have knowledge; justification is needed.

However, as plausible as it may seem, I think that this is not what Dewey has in mind. True, Dewey puts great emphasis on the importance of warrant or justification: the very idea of substituting knowledge with warranted assertion is intended to highlight the centrality of the justificatory relation between judgment and the course of action of which the former is the settled outcome. However, the function of justification or warrant is not that of putting the inquirer in the best position to assess the truth of a belief, but rather that of making it possible for a linguistic expression to be true. In this sense, warranted assertion does not coincide with truth, but it is that specific condition – related to the temporal and functional structure of inquiry – that makes truth possible.[39] As Dewey puts it, "truth and falsity are properties only of that subject-matter which is the *end*, the close, of the inquiry by means of which it is reached" (LW14, 176, emphasis in original).

Accordingly, if my reading is correct, Dewey's notion of warrant is remarkably different from the epistemological concept of justification. Broadly speaking, from an epistemological point of view, justification is an epistemological condition that has to do with the epistemic reasons that an agent has to hold a belief. Dewey's notion of warrant seems to work differently: it is less an epistemic condition than an

[39] In some other situations, Dewey is willing to take a step further and explicitly identify *true* with *verified*, as, for instance, in the following passage from *Reconstruction of Philosophy*: "In physical matters men have slowly grown accustomed in all specific beliefs to identifying the true with the verified. But they still hesitate to recognize the implication of this identification and to derive the definition of truth from it [...]. To generalize the recognition that the true means the verified and means nothing else places upon men the responsibility for surrendering political and moral dogmas, and subjecting to the test of consequences their most cherished prejudice" (MW12, 171). Clearly, from the standard epistemological point of view, warrant and verification – especially if verification is understood in experimental terms, as Dewey sometimes seems ready to acknowledged – are by no means synonyms. So, for instance, I am warranted in believing that yesterday Jones had toothache because I met him ten minutes ago and he told me so, even though I cannot verified it. In Dewey's own terms, nonetheless, such identification seems less problematic: inquiry only ends when the plan of action through which the indeterminate situation is reconstructed is tested and proves effective in achieving the goal that it purports to accomplish. According to this reading, warranted assertion would coincide with verified assertion. I am not sure, however, about the theoretical legitimacy of using the notion of verification to clarify that of warranted assertion. In the *Logic*, for instance, Dewey never uses – at least, to my knowledge – the notion of verification to clarify that of warrant.

activity, which is not completed until the final assertion is acted upon, by which it proves to be an effective tool for reconstructing the problematic situation that called out inquiry.

Now, as remarked above, when the conclusion of inquiry is reached, the situation is no more internally divided and conflictual: the significances of the objects of the situation hang together, and no break exists between significances and meanings. Dewey is clear on this point: "my position," he writes, "is that something of the order of a theory or hypothesis, a meaning entertained as a *possible significance* in some actual case, is demanded, if there is to be *warranted* assertibility in the case of a particular matter of fact" (LW14, 169).

This passage is of the utmost importance. On a general level, it formulates with the greatest clarity the kernel of what I call the semantic reading of inquiry – namely, the idea that inquiry is the process through which a meaning which is at first entertained as a *possible* significance becomes, when inquiry is ended, an actual significance which prompts overt action. Secondly, it stresses that, properly speaking, there is no knowledge outside and independently of inquiry, and that inquiry is closed – and 'knowledge' achieved – when meanings are finally used in actual courses of activity. In this sense, Dewey remarks, his is a correspondence theory of truth, provided that correspondence is understood in an operational way, as when it is said that a key corresponds to a lock or an answer adequately corresponds to a question. In this sense, correspondence – but it would be better to speak of 'corresponding' – is deprived of any metaphysical or epistemological connotations, thus becoming an experimentally detectable property of a particular form of behavior (Fesmire 2014, 103ff.). Finally, that passage shows that there is an intrinsic relation between the controlled construction and testing of meanings during inquiry and the justification or warrant of the final assertion.

So far so good. What has been shown thus far is that judgment is something which we are ready to take as a guide for action: when the assertion is warranted, we have warrant in moving forward. Thus, we stop reflecting – i.e., suspending overt action – and start modifying the objective conditions of the problematic situation.[40] I believe, however, that this set of assumptions may give raise to a serious difficulty for Dewey's logical theory, which revolves around the unrevisability of warranted assertion.

Take a very simple case of inquiry: for instance, a man who perceives a fire in a room, and searches to reach a safe place in the easiest way possible. Clearly, the only way to satisfy all the demands of the case is through reflection. The man in question may decide to jump out of the window without pondering his action; and it may even happen that he will land safely; however, unless he already knew that jumping out of *that* particular window is not dangerous – in which case there is no inquiry at all, but just habitual action – his action is not grounded, and, consequently, the conclusion of the activity is unwarranted. On the contrary, if the course of action is backed by a satisfactory analysis of data, as well as by an appropriate testing of the different

[40]I would like to thank David Hildebrand for helping me to clarify my thoughts on this point.

possible planes of action by which the problem can be handled, the conclusion thus reached is warranted, the judgment is true, and the final, determinate situation is ready to occur.

In cases like this, since there is no reason to defer the action any longer, the course of activity that the judgment prescribes is undertaken; as a consequence of that, the objective conditions of the situation are existentially modified. In doing so, the situation acquires new objective properties: those features that called out inquiry are transformed once for all, and thus are no longer available for further inspection. More radically, they could not, *in principle*, present themselves again. Indeed, Dewey is explicit in saying that the adaptation to the environment is not the restoration of a previous state of relative stability (LW12, 34,35; see also LW10, 19–20). Such a restoration would fly in the face of the transactional unity of organism and environment: even if it were possible to restore the same objective conditions, the agent would be nonetheless different since she would know how to (or how not to) behave in the new situation. Accordingly, the two situations are constitutively different, as are their conclusive phases.

My suggestion is that, in the case of common sense inquiries, the unrevisability of judgment not only is unproblematic, but also responds to an intuition that is worth preserving. Recall on which basis Dewey draws the distinction between scientific and common-sense activities: while the former are guided by the search of knowledge for the sake of knowledge, the latter are concerned with the practical affairs of everyday life. Common-sense inquiries are prompted by practical problems, and end when a solution is reached. In these cases, it seems reasonable to say that if the conclusion satisfies all the demands of the specific problem that called out reflection, there is no need to challenge or question the validity – better said, the truth – of the warrant assertion in which inquiry ends.

If the man who searches to get out from a room where fire broke out reaches a safe place without hurting himself, there seems to be no reason – aside from sceptical concerns, which are not legitimate within a pragmatist framework – to challenge the truth of the warranted assertion "the staircase near that door is the easiest and safest way to leave the room". Suppose that this solution answers perfectly – not in an absolute sense, but according to the standard and demands of the situation – to the specificities of the situation; it therefore corresponds to reality, and truth accrues on it. That linguistic expression has proved itself to be reliable in a course of activity; it is warranted in the sense of being verified as the proper answer to the particular case in question. Which does not mean that no other answer would be *theoretically* possible. The point is, rather, that if the solution is satisfactory, there is nothing more to say about it: in most of the cases, once the reconstructed situation is brought about, the objects of the situation acquire new significances, and the agent learns how to use them properly in future, similar situations. Accordingly, the unrevisability of warranted assertion is of a practical kind: it implies that the warranted assertions of previous inquiries are taken for granted and acted upon, without further reflection.

Scientific inquiries are much more difficult to accommodate within this framework of explanation. Indeed, in the case of scientific inquiries it seems natural to hold that their results are essentially open to further testing and modification. In the last analysis, scientific inquiries are concerned with the search of knowledge for the sake of knowledge, and they cannot stop before knowledge is achieved. Putting this point in terms of truth, there seems to be a strong intuition in support of the view that the fact that a certain assertion is reached as the warranted conclusion of a scientific process of inquiry does not imply that it is true. In other words, unless one is willing to endorse the view that what is currently believed about scientific objects is actually true – not *approximately* true, but absolutely true – she is compelled to acknowledge that fallibilism is the only option available on the market.[41] The strategy that has been pursued to assess the case of common-sense inquiries is, therefore, not accessible. Something different has to be attempted.

For the sake of honesty, I confess that I do not have a clearcut solution to this problem. I will advance two different answers that I think have some plausibility, but I admit that everything that I am going to say in this regard is quite speculative. I leave the whole question open for further discussion; in any case, I believe that this is an issue that any reconstruction of Dewey's philosophy of science should address.

The first proposal goes in the direction of stressing the normative value that Dewey attributes to the notion of inquiry. As said, no inquiry is closed until a successful, true conclusion is reached. Now, since it is widely agreed that scientific theories are, at best, approximatively true, it is quite straightforward to conclude that, properly speaking, no warranted scientific assertion has ever been made. According to this reading, all the scientific inquiries that have come to an end in the past or are still going on are to be understood as moments of a broader, open inquiry. True, such an inquiry has already brought about some intermediate partial judgments, as Dewey calls them; and it is also important to notice, as Dewey is ready to do, that "final judgment is constructed by a series of intermediate partial judgments" (LW12, 144). Nonetheless, the final judgment that will end inquiry and will reconstruct the problematic situation is still to come.

This reading is clearly close in spirit to Peirce's approach. Accordingly, it makes it easy to see on which basis Dewey accepts Peirce's *definition* of truth.[42] Moreover,

[41]A remark here is needed. Dewey's theory of inquiry is rightly considered as one of the prototypical forms of fallibilism. Obviously, I agree with that, and my insistence on the unrevisability of judgment does not purport to challenge the standard interpretation of Dewey's thought. As I read it, Dewey's fallibilism hinges on the idea that *propositions* – both universal and existential – are always revisable: they can be checked and modified in the course of inquiry, as well as when they are transferred from one inquiry to another. If my reading is correct, it is possible to combine fallibilism (about propositions) and unrevisability (about judgment).

[42]Here is what Dewey writes in this regard. "The best definition of truth from the logical standpoint which is known to me is that by Peirce: 'The opinion which is fated to be ultimately agreed to by all who investigate is what we mean by the truth, and the object represented in this opinion is the real.' *Op.cit.*, Vol. V, p. 268. A more complete (and more suggestive) statement is the following: 'Truth is that concordance of an abstract statement with the ideal limit towards which endless investigation would tend to bring scientific belief, which concordance the abstract statement may possess by virtue of the confession of its inaccuracy and one-sidedness, and this confession is an essential ingredient of truth.' (*Ibid.*, pp. 394–5.)" (LW12, 343).

a reading along this line has the merit of clarifying the function that the idealized notion of truth can play in Dewey's philosophy of science; it also accounts for his idea of the continuity of scientific inquiry. However, these theoretical advantages go hand in hand with some relevant difficulties. I focus on two key aspects. First of all, this reading seems to be overtly prescriptive and, to some extent, counterintuitive. Indeed, it is quite natural to use the term 'scientific inquiry' to refer to the various, different and independent activities that are and have been carried out by scientists; in this sense, the notion of scientific inquiry is understood as substantially equivalent to that of scientific practice. Now, if we accept this first proposal, it seems to me that we run the risk of loosing all the explanatory advantages related to the idea of pluralism of practices. In addition, it becomes difficult to understand who is the inquirer in charge of the broader inquiry of which the various, concrete scientific inquiries would be constitutive moments. One might suggest that the inquirer is the community of inquirers, but an answer along this line is not convincing; firstly, it is not clear how the normativity of the situation could exert its function of control and direction. Secondly, and most importantly, this reading tends to increase the difference between scientific and common-sense inquiries, thus being at odds with Dewey's idea of the "community of pattern in these two distinctive modes of inquiry" (LW12, 118).

The second proposal aims to correct the shortcomings of the first one, and it does so by focusing on the concreteness and individuality of *any* inquiry whatsoever. Contrary to the first proposal, this second one emphasizes the continuity between scientific and common-sense inquiries.[43] Accordingly, any scientific inquiry is to be understood as an autonomous and independent activity of problem finding and problem solving. Every scientific inquiry originates from a *specific* problem and searches for a *specific* solution: what is worth noting is that, according to this reading, the satisfactoriness of the solution depends both on the standards of success and the resources – conceptual as well as technological – that are available to the inquirer in the *particular* moment in which she carries out the inquiry. Clearly, the standards of success have not to be conceived of in a strongly constructivist fashion; there is no need to maintain that an inquiry responds exclusively to cultural or political influences; intuitively, the most important among the standards of success is that the world answers to the activities of the inquirer in the way that she expects it to do. More on this on the next chapter. For now it is enough to note that the

[43] As remarked in Chap. 1, this is a desideratum that Dewey believes a theory of inquiry should be able to satisfy. See, for instance, what he writes in the ninth chapter of *Experience and Nature*: "[i]t would then be seen that science is an art, that art is practice, and that the only distinction worth drawing is not between practice and theory, but between those modes of practice that are not intelligent, not inherently and immediately enjoyable, and those which are full of enjoyed meanings. When this perception dawns, it will be a commonplace that art – the mode of activity that is charged with meanings capable of immediately enjoyed possession – is the complete culmination of nature, and that 'science' is properly a handmaiden that conducts natural events to this happy issue" (LW1, 268–269).

highly contextual nature of these standards makes it possible for an assertion to be warranted without entailing its infallibility.

I prefer this second proposal.[44] It seems to me to be theoretically stronger. Not only it preserves fallibilism; it also encompasses some perspectivist insights that can be formulated in terms that could be palatable to contemporary philosophers of science working on the notion of scientific objectivity. In addition, it is faithful to some important Deweyan assumptions, such as the fact that for a certain activity to be an inquiry, the indeterminate situation must be so made as to contain within itself the elements for its solution.

As said above, I do not want to take stance on this issue: I am aware that this second proposal, too, suffers some difficulties. For instance, to name only one important example, it seems to imply that the identification between verification and truth that Dewey sometimes emphasizes has either to be abandoned or substantially revised (see, for instance, MW12, 171). I leave, therefore, the question open. In any case, I must confess that I have the impression that any account of Dewey's notion of warranted assertibility that aims to be sound from a textual point of view and theoretically consistent should be willing to acknowledge the possibility of dismissing or revising some relevant features of his theory of inquiry.

---

[44] More importantly, this seems to be Dewey's position too. In a letter to Kaufmann, Dewey clarifies what he means with the notion of warranted assertibility. He writes: "'warranted assertibility' as I understand it is that of a grea[t] specifiable temporal-spatial situation – not absolute. It is enough, speaking practically, that we do the best we can with the resources at hand to direct inquiry– (or knowing–) effectively. The positive postulate is that in so doing the 'methods' will improve and its so that their conclusions will be better warranted from the standpoint of assertibility. The critical postulate is that isolation depends upon acceptance in use of some relic of supernaturalism" (1945.02.26, John Dewey to Felix Kaufman).

Chapter 4
Construction

Abstract This chapter elaborates on the results reached in Chap. 2, and is devoted to the analysis of the functional elements through which scientific inquiry is carried out. Dewey is explicit that all the elements that are used in inquiry – and which he calls 'propositions' – are linguistic, and that taken together they contribute to the construction of the judgment that closes inquiry. I will reconstruct Dewey's account of evidence, a priori knowledge, and objectivity, and I will show how the latter paves the way for the formulation of a strong form of scientific pluralism.

Keywords Judgment · Synthetic · Copula · Predication · Existential and universal propositions · Evidence · A priori · Objectivity · Logical notion · Inferential stability · Pluralism

4.1 Introduction

In the previous chapter I have discussed inquiry at large. It is now time to take a step further, and turn attention from the temporal sequence of inquiry to its functional pattern. The goal of this chapter is to analyze the fundamental structure of the latter, and, in so doing, to lay out and articulate in detail what I have called the semantic interpretation of Dewey's theory of inquiry.

In what follows, I take it for granted that Dewey's theory of inquiry counts as a form of constructivism.[1] This should not be a matter of debate since Dewey is outspoken about the reconstructive function of inquiry, and it is hardly questionable that, all its peculiarities notwithstanding, re-construction is a form of construction.

[1]This is not a controversial assumption in Deweyan scholarship. See, for instance, the book *John Dewey Between Pragmatism and Constructivism*, in which Dewey's constructivism is analyzed from various different perspectives (philosophy of education, philosophy of technology, philosophy of culture) (Hickman and Reich 2009). As will be made clear in the following, the peculiarity of my approach resides exclusively in the logical tone that I attribute to Dewey's constructivism, not in the fact of understanding Dewey's theory of inquiry – and, consequently, his philosophy of science – as a form of constructivism.

R. Gronda, *Dewey's Philosophy of Science*, Synthese Library 421,
https://doi.org/10.1007/978-3-030-37562-1_4

The issue at stake is, therefore, to grasp which version of constructivism he advocates. Indeed, constructivism is a broad label which encompasses many different epistemological and ontological positions.[2] Some of them – those more radical, like the ones upheld by many contemporary sociologists of science – deny the very possibility of objective knowledge. Since knowledge is the result of a social process of negotiation, so their argument goes, its validity is a sociological and cultural affair, which lacks epistemic justification. Constructivism thus conceived is capable of supporting a communitarian form of relativism.

Dewey's constructivism is not of this blend. First of all, it is a *logical* constructivism: that means that the kind of constructivism that Dewey advocates is subjected to the normative guidance exerted on inquiry by the situation that prompts reflection. Moreover, because of its realistic spirit, Dewey's constructivism does not support relativistic conclusions: since the goal of inquiry is to transform the objective conditions of the situation from which inquiry stems, knowledge – Dewey would say, warranted assertion – cannot be depicted as the result of a process of social negotiation. It necessarily implies some sort of manipulation of the features of the environment under investigation. Similarly, Dewey rejects any position that starts from the assumption that a subject exists who projects his categories on a raw material. The primacy that he attributes to the notion of situation is functional, among the other things, to getting rid of the idea that knowledge can be understood in epistemological terms.

As said, Dewey's constructivism hinges upon the concept of reconstruction.[3] But what is reconstructed in the course of inquiry? I have repeatedly noticed that my interpretation of Dewey's theory of inquiry is semantic: consequently, I hold that what is reconstructed in and through inquiry is the significance of some objects of the indeterminate situation. According to the semantic interpretation that I will articulate in the following pages, the goal of inquiry is to transform the objective conditions of an indeterminate situation so as to make it possible for the objects to embody as their proper significances the meanings of the symbols that are predicated of them.[4]

---

[2]The bibliography on constructivism is enormous. The best attempt to characterize and distinguish the various forms of constructivism remains (Hacking 2000). See also Kukla (2000) for a detailed analysis of the debate on constructivism in philosophy of science.

[3]The theoretical relevance of the distinction between construction and reconstruction, as well as the realistic implications of the latter notion, is highlighted by Godfrey-Smith in the article *Dewey and the Question of Realism*. See, in particular, Godfrey-Smith (2013a, 77).

[4]An interpretation along this line may seem to stand in tension with the naturalistic and realistic commitment attributed to Dewey's perspective. However, this conflict is mainly illusory, and depends on what I take to be a too narrow interpretation of Dewey's notion of language. As has been highlighted in Chap. 2, Dewey's semantic theory – that is, Dewey's theory of meaning and significance – is also framed within a naturalistic paradigm. Meaning and significance, symbols and signs, are modes of cultural and biological behavior, rather than intermediaries between mind and world. Consequently, the two aspects of inquiry – objective modification of the environment and semantic reconstruction of the significances of the objects of the situation – can be seen as two different ways of describing the same phenomenon, and are, therefore, compliant one with the other. I give pride of place to the semantic reading on the basis of Dewey's insistence on the linguistic nature of inquiry – see below Sect. 4.2.

In the four sections that make up the present chapter I will present the main features of Dewey's semantic constructivism. In Sect. 4.2, I will lay out Dewey's constructivist account of inquiry. I have provocatively framed the whole question in Kantian terms, asking how synthetic judgments can be possible. I think that the comparison with Kant is useful to shed light on the distinctive features of Dewey's position, and, in particular, on his idea of empirical evidence and theoretical hypothesis as co-constructed in the course of inquiry.

The three remaining sections are devoted to analyzing – from the functional perspective that is distinctive of Dewey's approach – three notions that are central to any philosophy of science. In Sect. 4.3, I will deal with Dewey's constructivist conception of evidence, and I will argue that the latter is rich enough and powerful enough to satisfy the requirements that a good theory of evidence is expected to meet. In Sect. 4.4, I will address one of the most problematic aspects of Dewey's philosophy of science, namely, his account of the function of the a priori in science. Finally, in Sect. 4.5, I will discuss Dewey's functional theory of objectivity, and I will show that it supports a constructivist account of the nature of scientific objects which, in its turn, leads to a radical form of scientific pluralism.

4.2 How Are Synthetic Judgments Possible?

As hinted at above, in accordance with my semantic account of Dewey's theory of inquiry I take the goal of the latter to be to explain how synthetic judgments are possible.[5] This way of framing the issue may look surprising, since Dewey is usually associated with Hegel and the idealistic tradition rather than with Kant.[6] However, there is plenty of textual evidence to prove that Dewey's logical theory is in strong continuity with the Kantian project, to such an extent that the former can be read as an attempt to revise and reformulate, in a functionalist way, the latter.

Take, for instance, the *Logic*. In this text, Dewey refers to Kant's famous dictum that "[t]houghts without content are empty, intuitions without concepts are blind" (Kant 2016, A51/B75), and he explicitly acknowledges that Kant's is the proper

[5]Since in his works Dewey is almost exclusively concerned with common-sense and scientific inquiries, I could have said that the goal of his theory of inquiry is that of explaining how synthetic judgments *a posteriori* are possible. Now, I think that this (narrower) reading is correct, and I am ready to accept it. However, in order to properly argue for it, I should reconstruct Dewey's philosophy of mathematics from his sketchy and sparse remarks on that topic, and then assess whether, according to Dewey, mathematical judgments are analytic or synthetic. This is a highly interesting issue that deserves to be carefully investigated, but which I cannot pursue here. For this reason, I have preferred to speak of synthetic judgments in general, leaving the question of whether there is any real difference between synthetic judgments and synthetic judgments *a posteriori* open for further discussion.

[6]A remarkable exception is Seigfried (1993).

way of framing the problem on knowledge. He writes that "[t]he Kantian formula that apart from each other 'perceptions are blind and conceptions empty' marks a profound logical insight" (LW12, 114). Kant is right, Dewey argues, in stressing the fact that knowledge is necessarily constituted of two different elements – the empirical evidence and the theoretical framework in which the empirical evidence is interpreted.

Accordingly, Dewey agrees with Kant that knowledge is made possible by the interaction between an empirical and a conceptual component. He rejects, however, Kant's explanation of how such interaction takes place. As Dewey reads it, Kant maintains that those two elements are provided by two distinct faculties of our mind, sensibility and understanding; and he also maintains that it is thanks to the transcendental unity of apperception that all of the manifold given in an intuition is united in the concept of the object.

Dewey rejects the Kantian account for two different reasons. On the one hand, he argues that intuitions and concepts should not be understood as stemming from two independent faculties of knowledge. Dewey believes Kant's terminology to be misleading, in that it uses notions drawn from the psychology of faculty. Since that psychology depicts human mind as being made of different and independent faculties, the parcelization of mental activity to which it leads prevents from seeing the original unity of human *behavior*, as well as from understanding – on a distinctly philosophical level – that a satisfactory account of human mind should start from the assumption that the latter is a natural phenomenon which has to be investigated experimentally. In addition, the choice of that language prevents from realizing that the two components of knowledge should be taken as linguistic rather than as psychological or mental entities.

On the other hand, Dewey deems Kant's account as explanatory unsatisfactory. Dewey's argument goes as follows: if intuitions and concepts – or, in different words closer to Dewey's intentions, evidence and hypotheses – are assumed to be distinct and independent elements of knowledge, any possible attempt to synthesize them is either bound to fail, with the two remaining as distinct after the process of synthesis as they were at the outset, or to be unexplainable from an experimental perspective.

Clearly, these two criticisms are closely interrelated: the acknowledgment of the functional nature of the distinction between 'intuition' and 'concept' is a necessary precondition for a proper solution to the problem of synthesis. Accordingly, the two objections can be read as two parts of the same argument, which aims to provide a clarification of how synthetic judgments are possible. In the following pages, I will therefore proceed as follows. First of all, I will discuss Dewey's rejection of the Kantian idea that intuitions and concepts stem from two independent faculties, and I will lay out his linguistic account of evidence (intuition) and hypothesis (concept). Then, I will turn the attention to the second issue. I formulate it as follows: how can the idea of a functional and linguistic difference between evidence and hypothesis lead to a satisfactory account of their synthesis in judgment? The answer to this question will lead to highlighting the *semantic* primacy of inquiry over its components.

As many other philosophers working in the post-Kantian tradition, Dewey's aim is to correct Kant's dualistic approach in a way that makes it possible to show that concepts and intuitions are two essentially interrelated elements, and, consequently, that knowledge is possible precisely because they do not need to be synthesized. Far from being different and independent one from the other, they are correlative functions that cooperate in the construction of a judgment: as such, as two correlative functions, they are attuned one to the other from the very beginning of the process of inquiry.

This is how Dewey states the issue at stake:

> [Kant's] insight, however, was radically distorted because perceptual and conceptual contents were supposed to originate from different sources and thus required a third activity, that of synthetic understanding, to bring them together. In logical fact, perceptual and conceptual materials are instituted in functional correlativity with each other, in such a manner that the former locates and describes the problem while the latter represents a possible method of solution (LW12, 114–115).

Dewey's position is clear: he accepts the Kantian insight of the indispensability of evidence and theory, while refuting the thesis of their independence. But what does this mean, concretely? What are the philosophical consequences of such approach? And, more precisely, what are the consequences on the understanding of the possibility of synthetic judgments?

Dewey's response runs as follows. One of the consequences of Kant's dualistic approach – at least, as Dewey understands it – is that it supports a form of semantic 'atomism', so to say, that is incompatible with the holistic approach that Dewey endorses – at various different levels. Indeed, if concepts and intuitions are taken to stem from two different and independent sources, their logical import is – at least partially and to a certain extent – determined irrespectively of the function they play within inquiry. Clearly, in conformity with Dewey's usage, with the expression 'logical import', or 'logical value', I mean to refer to the specific function that a certain element plays in inquiry.

In light of this, the disagreement between Kant (or Dewey's Kant) and Dewey can be reformulated as follows: do intuitions and concepts – evidence and scientific hypotheses – have a content before and independently of their contextual use in the specific inquiry that is carried out in response to a specific problem?

Dewey's answer to that question is radical: he not only maintains that nothing has logical value outside inquiry; he also argues that it is impossible to understand the logical import of any element used in inquiry without referring to the function that it plays in that specific course of activity. In other words, there is no objective, independent sign of logical value.

The first thesis – no logical value outside inquiry – is neither surprising nor problematic. At the end of the day, it is nothing but a restatement of Dewey's view of the normative primacy of the situation over its elements. It is also in strong agreement with Dewey's overall rejection of the Myth of the Given. In this case, what is rejected is the idea that there might be some sort of givenness – logical givenness, so to say – that puts external constraints on inquiry. Rather the contrary, inquiry is normatively original.

The second thesis – no objective sign of logical value – is more interesting for my present purposes. It is here that the difference between Dewey and Kant's approaches becomes clear. According to the Kantian framework, intuitions and concepts have distinctive features, which enable to distinguish their contribution in the constitution of knowledge. Kant is explicit on this point: "[t]he understanding is not capable of intuiting anything, and the senses are not capable of thinking anything"; for this reason, he concludes, "these two faculties or capacities cannot exchange their functions" (Kant 2016, A51/B75).

Dewey agrees with Kant that evidence and hypothesis perform two different functions, but he does not endorse the view that such a difference in functions has to be traced back to some kind of underlying, substantial difference between intuitions and concepts. Stated in other words, there is no need of a metaphysical backup in order to account for the functional distinction between the two components of knowledge. Evidence and hypothesis are distinguished exclusively on the basis of their function within a course of inquiry.

Clearly, Dewey is trying here to sever logic from ontology. Far less clear is how he thinks he can achieve that goal. What does it mean, concretely, that the difference between evidence and hypothesis has to be cashed out in purely functional terms? What theoretical resources are needed to consistently formulate that position?

Dewey's strategy revolves around two pivotal assumptions. The first one follows directly from his refusal of Kant's 'psychological' vocabulary, and its substitution with a linguistically-based perspective. Contrary to Kant, who acknowledges some sort of relation between the distinctive qualities of intuitions and concepts, *qua* intuitions and concepts, and their function in knowledge, Dewey maintains that both evidence and hypothesis are linguistic. As linguistic entities, they are all on the same level, so to say: they are all of the same kind. In so doing, by taking a linguistic turn, a sound functionalist approach is made possible.

Before moving on to discussing the other assumption, let's examine this one for a moment. Such examination is needed since it is hardly deniable that some of the consequences of this linguistic turn may easily look paradoxical to our contemporary eye. Take, for instance, the *scientific* notions of evidence and hypothesis. Clearly, though it has been a highly debated issue in the last 50 years, it is not impossible to argue for the linguistic nature of scientific theories. It is much less easy to understand how empirical evidence can be said to be linguistic. One is naturally led to think that Dewey somehow confuses empirical evidence – say, the event of a litmus paper turning red after being immersed in a liquid – with its linguistic expression. However, if this were actually what Dewey had in mind, the thesis would be trivial, and his whole argument would lack any real force. For the same reason, indeed, it could be said that hunger is a linguistic phenomenon since I can utter the sentence "I am hungry" when I find myself in that state. But this is clearly absurd.

So, is this really the whole of Dewey's argument? I believe that Dewey's position is more refined than it may seem at first glance. As I read it, Dewey does not confuse empirical evidence with its linguistic formulation; rather, he deliberately privileges the latter over the former. The issue is, therefore, to understand why he decides to do so.

The reason of his choice is related to Dewey's idea of linguistic behavior. As has been highlighted in Chap. 2, the fundamental function of language is that of delaying immediate action, thus enabling the agent to carry out complex activities. It is precisely the capacity of language to block overt action that accounts for Dewey's insistence on the linguistic character of evidence. The search for evidence – or observation – is, properly speaking, an activity that takes place within inquiry. In this sense, it is an operation whose ultimate function is that of contributing to the construction of a new and more reliable habit of behavior, which is better fitted to the existing environmental conditions. In order to be effective as an element of inquiry, evidence must not be used as a stimulus for overt action. Accordingly, evidence *must* be linguistic since it must not issue in an overt action.

Dewey formulates this point in the following way. I quote the passage in its entirety since it is relevant to the ongoing discussion:

> It is obvious [...] that a possible mode of solution must be earned in symbolic form since it is a possibility, not an assured present existence. Observed facts, on the other hand, are existentially present. It might seem therefore, that symbols are not required for referring to them. But if they are not carried and treated by means of symbols, they lose their provisional character, and in losing this character they are categorically asserted and inquiry comes to an end. The carrying on of inquiry requires that the facts be taken as *re*presentative and not just as *pre*-sented. This demand is met by formulating them in propositions – that is, by means of symbols. Unless they are so represented they relapse into the total qualitative situation (LW12, 118).

The latter remarks pave the way for the analysis of the other assumption of Dewey's functionalist proposal. As has been repeatedly remarked, Dewey's philosophy of language is strongly contextualist. Recall Dewey's example of the different meanings associated with the word "meza" (see above 2.3). The meaning of a symbol depends on its use in a specific course of action, and is left completely undetermined when the contextual information is lacking. Now, as linguistic entities, both evidence and hypothesis are contextual. This means that there is no objective linguistic element that can be used to distinguish the one from the other.

Dewey is explicit on this point. Take, for instance, a sentence like "All men are mortals". Dewey argues that it is not possible to tell whether it is an existential proposition providing the evidence of the case – the fact that all the human beings who have ever lived have died – or an universal proposition which formulates the conditions of use of that notion, to the effect that when someone is said to be a human being, it can also be said that she is mortal. Since any distinction between these two propositions is contextual, it has to be achieved in terms of the functions that they perform within inquiry. Dewey's contextualism and his functionalism go therefore hand in hand. Dewey expresses that connection with the following words:

> every term (meaning) is what it is in virtue of its membership in a proposition (its relation to another term), and every proposition in turn is what it is in virtue of its membership in either the set of ordered propositions that ground inference or in the series of propositions that constitute discourse. It follows from this position that the logical content and force of terms and propositions are ultimately determined by their place in the set of propositions found in either inference or discourse (LW12, 327).

Let's take stock. Up to now I have laid out the argument that Dewey puts forth in support of his functionalist account of the two components of knowledge. Contrary to Kant, Dewey holds that the only difference between intuitions and concepts is functional, and that evidence and hypothesis are linguistic entities, whose logical value is entirely determined by the contribution they make to the reconstruction of the problematic situation that originated inquiry.

But this is just the first step towards the solution of the problem of the possibility of synthetic judgments. As will be remembered, Dewey's objection to Kant is twofold: firstly, he remarks that intuitions and concepts should not be understood as stemming from two independent faculties of knowledge; and, secondly, he argues that on that basis their synthesis would be unexplainable. So, the next step is to show how the idea of a merely functional difference between evidence and hypothesis can lead to a satisfactory account of their synthesis in judgment.

A preliminary consideration is necessary here. As has been remarked in the previous chapter, Dewey distinguishes – theoretically as well as terminologically – between those sentences that are used within the course of inquiry, which are only provisional and affirmed as far as the inquiry goes, and the unique sentence which concludes inquiry. The former are called propositions, while the latter is called judgment.

Now, as said, judgment and proposition cannot be distinguished grammatically, the ground for their distinction being merely functional. In order to illustrate such a distinction, consider, for instance, a trial. In a trial, the very same sentence – say, "the suspect A has been found with blood still on his hands and clothes, and DNA testing has verified that the blood is that of the victim" – can be uttered by the scientific expert in her medical report and by the judge while reading the sentence. The two utterances, for the simple fact of being uttered in two different moments of inquiry, have different logical value: in the first case, the sentence states a *possible* fact, whose legal import is to be assessed; in the second case, it states the *actual* fact that the suspect A committed the murder. Or take a scientific observational report such as that neutrinos travel faster than light. Some years ago, it was announced that the experiment OPERA had shown that a beam of neutrinos coming from CERN in Ginevra were registered arriving 60 nanoseconds faster than the speed of light allows. Clearly, the import of such report is different whether it is taken as a set of observational data that stand in need of account, as well as assessment, or as an empirical finding. Only in the latter case the report has existential bearing: it predicates a certain property of a certain subject.

The point which is relevant for my purposes is that, precisely because of their provisional nature, Dewey does not see any problem with the kind of synthesis that makes *propositions* possible. Since a proposition does not properly attribute a predicate to the subject – if it did, it would be the conclusion of inquiry – the predication is not real, but is only tentatively made. With this distinction in mind, Dewey neatly separates affirmation from assertion, thus highlighting the difference existing between "the logical status of intermediated subject-matters that are taken for use in connection with what they may lead to as means" (affirmation) and "subject-matter which has been prepared to be final" (assertion) (LW12, 123).

The rationale behind this view is that the predication made in a proposition does not purport to say something about the world. The aim of a proposition – of *any* proposition, even those that provide empirical evidence – is not that of asserting how things really are. Obviously, such remark is to be understood in light of Dewey's practice-oriented approach. That statement does not mean that propositions are ungrounded; nor that they are *merely* hypothetical – though a good illustration of the kind of predication which is distinctive of propositions may be the procedure of *reductio ad absurdum*, in which the proposition to be disproved is affirmed without being asserted. Rather, it means that their function consists in being employed as means to reconstruct a problematic situation. As means, their entire and exclusive logical value is that of being effective in conducting to a successful solution. It is for this reason that Dewey states that propositions – existential as well as universal – are neither true nor false:

> According to the position here taken, propositions are to be differentiated and identified on the ground of the function of their contents as *means*, procedural and material, further distinctions of forms of propositions being instituted on the ground of the special ways in which their respective characteristic subject-matters function as means [. . .]. [A]t this point it is pertinent to note that, since means as such are neither true nor false, truth-falsity is *not* a property of propositions. Means are either effective or ineffective; pertinent or irrelevant; wasteful or economical, the criterion for the difference being found in the consequences with which they are connected as means (LW12, 287).

An inquirer may be convinced that a certain proposition is so reliable that she is ready to take it as true. Nonetheless, if that proposition is used *as* a proposition in inquiry, it cannot be a truth-bearer, by definition. Indeed, true and false are properties of judgments, not of propositions. That does not mean, however, that propositions are epistemically idle. The idea of propositions as means for reconstructing a problematic situation is heuristically fruitful in that it paves the way for a different conception of their epistemic value: truth and falsity do not exhaust the scope of what has epistemic value. Now, a means can be more or less effective in leading to the realization of the goal that it purports to achieve, and part of the effort of inquiry – especially in the case of scientific research, which reaches a higher degree of internal complexity – consists precisely in the assessment of the validity of propositions as means for an end-in-view. The latter is an epistemic work, which has a distinctively progressive character.

Dewey formulates this point by remarking that, during the course of inquiry, those propositions that are judged to be reliable are subtracted from investigation, and directly acted upon. Consequently, their mode of predication changes from affirmation to assertion: they stop being held provisionally, and are accepted as tools of action, either as providing true and reliable empirical evidence or as stating true and reliable laws. Obviously, since they are asserted as *tools* or *means* for reconstructing the problematic situation that called out reflective activity, they are still revisable in the course of further inquiry: their logical value – their capacity to lead the present inquiry to its successful conclusion – is open to modification and revaluation. In this sense, Dewey calls those judgments which intervene in inquiry, and are the settlements through which the final settlement is brought about, "ad-judgments".

We are now in a better position to address the original problem – namely, how the idea of functional difference between evidence and hypothesis can lead to a satisfactory account of their synthesis in the judgment that ends inquiry. Dewey's solution is composed of three steps.

First of all, Dewey acknowledges that it is the same material that takes different forms throughout the course of inquiry. Consequently, it has to be distinguished, both conceptually and terminologically:

> As *undergoing* inquiry, the material has a different logical import from that which it has as the *outcome* of inquiry. In its first capacity and status, it will be called by the general name *subject-matter*. When it is necessary to refer to subject-matter in the context of either observation or ideation, the name *content* will be used, and, particularly on account of its *representative* character, content of propositions. // The name *objects* will be reserved for subject-matter so far as it has been produced and ordered in settled form by means of inquiry; proleptically, objects are the *objectives* of inquiry (LW12, 122).

The continuity of inquiry reverberates in the continuity of its subject matter. This means that there is no break between content, which is the subject-matter of propositions, and object, which is the subject-matter of judgment. The latter emerges out of the former, while, at the same time, constituting its articulation and refinement. In Dewey's technical vocabulary, an object is the form that accrues on the subject-matter of the situation, thanks to which what has been settled by and through inquiry is at disposal of the inquirer for her successive investigations. Accordingly, the transformation of the contents of the propositions into the objects of judgment can be accounted for in terms of the temporal sequence of inquiry.

The second step of the argument has to do with the process of co-construction of evidence and hypothesis. Dewey neatly distinguishes existential propositions from universal propositions. As said, existential propositions provide empirical evidence, while universal statements formulate the operations to be carried out on the existential material. The same insight can be conveyed in a slightly different form, by focusing on the different functions that the two kinds of propositions perform in relation to the construction of the final judgment. In this sense, existential propositions constitute the subject-content of the judgment, while universal propositions provide its predicate-content.[7]

Subject and predicate content, therefore, co-evolve out of the problematic situation that originated inquiry. The process of co-construction of existential and universal propositions is internally articulated, since the functions that they perform in the context of inquiry are structurally diversified. Nonetheless, being two moments of the very same process, their functions are essentially interwoven:

[7]During inquiry, the predicate content is attributed only tentatively to the subject, with the aim of accounting for the evidence provided by existential propositions. At the same time, the consequences that can be logically derived from the predicate content are sought after; if they are actually discovered or produced, they provide support to the hypothesis that was advanced to account for the data available at that moment.

the logical import of evidence is that of paving the way for the formulation of a hypothesis that could succeed in solving the problem at stake, while the logical import of hypothesis is that of directing some experimental operations that bring about new and better evidence, which, in its turn, confirms or raises a challenge to the hypothesis.

Expressed in semantic terms, the process of co-construction of evidence and hypothesis can be described as follows. The function of existential propositions is to assess the significance of those elements of the environment that are thought – even though only provisionally – to be relevant to the definition and solution of the problem. The notion of significance refers to the objective capacity of a thing to act as a sign of something else. Therefore, existential propositions state the potentialities of certain elements, on the assumption that those *potentialities* can be of help in reconstructing the problematic situation.

Universal propositions, on the contrary, formulate general modes of action that the inquirer can decide whether or not to carry out. Consequently, if existential propositions state the potentialities of the elements of the environment that are supposed to be relevant to inquiry, universal propositions state possible ways of acting, which are expressed in if-then propositions. Indeed, when a general mode of action is inhibited from direct manifestation and comes to be embodied in a symbol, it has to be treated as a *possibility*.

Here is how Dewey formulates that distinction:

> *Potentialities* are to be distinguished from abstract *possibilities*. The former are existential "powers" that are actualized under given conditions of existential interaction. Possibility, on the other hand, is a matter of an operation as such – it is operability. It is existentially actualized only when the operation is performed not with or upon symbols but upon existences (LW12, 228–289).[8]

[8]The functions of singling out potentialities and formulating possibilities are intrinsically different, and their difference is a difference in degree of acceptance as well as a difference in generality. It is a difference in generality because, contrary to existential propositions, universal propositions do not refer to any actual operation. An if-then proposition merely states a connection between a set of operations and its possible consequences, and the validity of the proposition depends on whether or not that connection is discovered to hold. On the contrary, by attributing certain powers to a certain thing, an existential proposition refers to specific operations that are to be performed on that particular object under some specific conditions. It is a difference in degree of acceptance because, contrary to universal propositions, existential propositions commit the inquirer to the reality of the connection between operations and their consequences. This white and granoulous substance has the power to solve in water: accordingly, the content of this existential proposition attributes the dispositional property 'soluble' to this table of sugar. So, though the corresponding operations are not actually performed, the existential power that is expected to be triggered by those operations is attributed to the object. The inquirer 'knows' that if she puts this table of sugar in a glass of water, it will solve. Similarly, the inquirer 'knows' that if she bombards the nucleus of an uranium-235 atom with neutrons, a nuclear fission may follow. The existential proposition "uranium-235 is fissionable" states a real potentiality of the object on which the inquirer can count in the course of further inquiry. On the contrary, the content of an universal proposition is not attributed to the object referred to by existential propositions as one of its properties. Dewey is explicit on this point: "the predicational content [...] is related to the factual content, that is, the subject, as the possible to the actual" (LW12, 134). Being in the mode of possibility – Dewey

Before turning attention to the third and final step of Dewey's solution to the problem of synthesis, let's draw a preliminary conclusion from all these remarks. According to my semantic interpretation, inquiry is to be understood as that process whose goal is to turn the possibilities expressed by universal propositions into potentialities that can be directly acted upon. Synthetic judgment – the outcome of inquiry – is the sentence in which such transformation is completed: the meaning stated by the predicate is attributed as a property of the significance stated by the subject. More precisely, judgment is that sentence in which the possibilities expressed by the manifold universal propositions used during inquiry precipitate in the predicate, while the potentialities expressed by the manifold existential propositions used during inquiry converge in the subject. When a judgment is formulated, reflection ends and overt action takes place.

Now, synthetic judgment is made possible by the fact that meanings (universal propositions) and significances (existential propositions) are progressively attuned to one another through and by the course of inquiry. As said, this amounts to saying that, during inquiry, meanings and significance are co-constructed. This is usually taken to mean that universal propositions are suggested by empirical evidence, in a way that is substantially homologous to Peirce's abductive reasoning. This reading is obviously correct. But that thesis also means – as Dewey explicitly acknowledges – that "[t]he significance of factual material is fixed by the rules of the existing [theoretical or scientific] system" (LW12, 124). In the case of a trial, which is Dewey's favorite example, it is the juridical system that fixes the significance of the factual material. In the case of a scientific experiment, it is the best theory available to the inquirer.[9]

The idea of co-construction of empirical evidence and theoretical hypothesis enables Dewey to explain how two elements that are functionally different can

speaks of "operability" to stress the fact that they are purely possible operations – the content of the universal proposition can be attributed to the subject without contradiction. However, that does not mean that for the mere fact of being a possible property it is thereby actual. The passage from possibility to actuality does not come for free; it requires some operation of an experimental kind, by which it is ascertained whether the object responds in the expected way to the actions made by the inquirer under the guidance of the hypothesis that constitutes the content of the universal proposition.

[9]Theories, therefore, construct the significances of the evidential data. Consequently, not simply the meanings that are advanced as possible ways to solve the problematic situation are tentative; the significance of the factual material is also held provisionally. Clearly, this result is in agreement with Dewey's theory of the linguistic nature of propositions. It is worth noting that factual material is provisional in two senses: it is provisional in the sense that it is not clear whether or not it is relevant for the case under discussion; and it is provisional in the sense that it is not clear whether or not it supports the interpretation provided by the meanings that constitute the hypothesis under investigation. As Dewey repeatedly remarks, when these two features of evidence are clarified, the validity of the hypothesis is also established, and inquiry reaches a conclusion. In the negative case, evidence and hypothesis turn out not to be relevant to the specific problem at stake; a (negative) adjudgment closes this line of investigation, and demands a different approach. In the positive case, the available evidence confirms and supports the hypothesis, which indicates which action to undertake; a judgment is eventually constructed, and inquiry ends.

be unified in a synthetic judgment. What remains to be done is to analyze the notion of predication that makes synthetic judgment possible. Indeed, the fact that subject and predicate coevolve within inquiry does not in itself account for the *act* of their predication. Contrary to Fregean tradition, which explains the unity of the proposition in terms of the saturation of a concept by an object, Dewey adheres to the classical paradigm, to the effect that the unity of the proposition is due to the unifying function of the copula. So, in the case of synthetic judgments the problem of understanding how the copula constitutes the unity of proposition can be formulated as follows: on which *ground* does the copula unify subject and object?

To my knowledge, Dewey's most extensive discussion of this issue is contained in *Qualitative Thought*. In that article, Dewey articulates with the greatest clarity what he calls "the sceptical difficulty" about the possibility of synthetic judgments. As he remarks, "[t]he sceptical difficulty may be summed up in the statement that predication is either tautological and so meaningless, or else falsifying or at least arbitrary" (LW5, 252).

The argument runs as follows. Take a sentence P, say, "that thing is sweet". Now, consider the relation between the subject and the predicate-content. The relation can be of two kinds. In one case, the content of 'sweet' already qualifies the content of 'that thing'. Accordingly, Dewey remarks, "the predication is analytic in the Kantian sense" (LW5, 252). This consideration is troublesome since, according to Kant, analytic judgments are universal and necessary, which is clearly not the case here. In addition, since the Kantian distinction between analytic and synthetic judgments has to do with the different kinds of justification that can be provided, an existential judgment cannot be assessed as true or false on the basis of the principle of contradiction. In any case, it is clear what Dewey has in mind. If the predicate-content is already included in the subject-content, from the point of view of the process of inquiry, it is fair to say that the predication does not add anything new to the content of the subject. Accordingly, it is tautological.

The second case is much less straightforward. First of all, Dewey is not explicit about the kind of problem that is at stake in the case of "real" synthetic judgments. He seems to oscillate between two different options: in one case, the predication is said to be falsifying; in the other, it is charged with being arbitrary. In both cases a difficulty arises, but its reasons are different.

Let's deal with the first possibility first. If the predication is falsifying, it seems fair to conclude that Dewey believes that a contradiction immediately arises. Indeed, the mere fact of stating the judgment implies that what is judged is false. Unfortunately, Dewey does not explain how this contradiction is supposed to arise, so that we can only advance hypotheses. My suggestion is to understand that contradiction in terms of the simultaneous affirmation and negation of the identity of subject and predicate contents. Being a judgment – take, for the sake of simplicity, an affirmative judgment – the predicate content is said to be identical, *to a certain extent*, with the subject content. At the same time, however, being a *synthetic* judgment, the two contents must not be identical, and nonetheless they must be related. In any case, since Dewey does not tell anything explicit about this point, my suggestion is tentative at best; other options may be available.

What about the other possibility, that of arbitrariness? Here things are far more interesting for my purposes. Indeed, the fact that predication is arbitrary does not seem to entail contradiction – neither directly nor indirectly. If so, however, this second horn is constitutively different from the first one. What is at stake here is not the allegedly contradictory nature of judgment, but rather the fact that the arbitrariness of judgment entails its lack of grounding and, consequently, of normative guidance. "[I]f 'sweet' does not already qualify 'that thing'," Dewey remarks, "what ground is there for tacking it on?" (LW5, 252). The problem of arbitrariness can be, therefore, formulated in terms of the incapacity of judgment to provide its own ground for predication.

If my reading is correct, the arbitrariness of judgment is to be taken as evidence of the fact that judgment – synthetic judgment – cannot be logically autonomous. So, the problem with judgment turns out to be that it cannot account for the predication that constitutes its unity. Considered as an autonomous entity, judgment lacks the expressive and normative resources to justify its own ground. Take again the judgment "that thing is sweet". There is nothing in this judgment which grounds that particular predication: why should that thing be qualified as sweet rather than, say, white, or granulose? In this sense the predication is left unexplained: many different predicates could be attributed to the subject. However, not every predication is equally satisfactory. Recall that judgment is the conclusion of a process of inquiry, thanks to which the problematic situation is reconstructed. Consequently, the arbitrariness of predication amounts to a denial of the very possibility of inquiry.

The way out from that difficulty – and this is the third and final step of Dewey's argument – is to acknowledge that judgment cannot be severed from the broader context of inquiry from which it originates. Dewey is clear on this point:

> The copula in a judgment, in distinction from the term of formal relation, expresses, accordingly, the actual transformation of the subject-matter of an indeterminate situation into a determinate one. So far is the copula from being an isolable constituent that it might be regarded as what sets the subject-and-predicate-contents at work executing their functions in relation to one another (LW12, 138).

For a judgment to be possible, it has to be grounded; and for a judgment to be grounded, it has to be a part of a greater whole – obviously, this whole is inquiry, which constitutes, therefore, the basic semantic unit. On the contrary, if judgment is isolated from the whole of inquiry, its predication becomes arbitrary: as a consequence of that, judgment is ungrounded, and inquiry lacks its normative guidance.

Now, if the primacy of inquiry is fully acknowledged, the problem of the ground of predication can be reformulated in wholly different terms. More precisely, within Dewey's practice-oriented framework, the synthetic function of the copula in a judgment can be explained in a way that refers exclusively to the overt activities performed by the inquirer. The copula in a judgment can be defined as that particular action that satisfies the specific need of the problematic situation that prompted inquiry, thus closing the whole course of reflective action. Or, stated in other words, the copula provides the unity of the synthetic judgment in that it is the act through which a *real* connection is established between the subject and the predicate content. The connection between the subject and the predicate content is now a fact of the world: the meanings expressed by the predicate become part of the significances expressed by the subject of judgment.

With this final move, Dewey believes that the problem of understanding how synthetic judgments are possible is solved. Before moving on to discuss Dewey's functional account of the notions of evidence, a priori, and objectivity, however, it may be useful to recapitulate the conclusions reached in this section. (1) Synthetic judgment is the outcome of a process of inquiry, which provides the normative guidance that is necessary in order for predication not to be arbitrary. (2) The ground of predication is the specific problem that called out inquiry. (3) The final determinate situation is that phase of inquiry in which the process of construction of objectivity comes to its end, and the significance of the object is enriched with the meanings expressed by the predicate of the judgment. (4) The process of construction is made possible by the co-construction and co-determination of significance (existential propositions) and meaning (universal propositions). (5) Existential and universal propositions can be distinguished only on a purely functional ground, in relation to the role they play in transforming the indeterminate situation firstly into a problematic situation, and then into a determinate and harmonic one.

4.3 The Construction of Evidence

In the previous section I have argued that Dewey's theory of inquiry can be reformulated as an attempt to understand how synthetic judgments are possible. Part of the discussion has centered on the function of existential propositions in inquiry: it has been remarked that their function is that of determining the problem involved in an indeterminate situation, in order to set the conditions for a controlled modification of the environment. Existential propositions have existential import in that they refer to specific spatio-temporal configurations: they pick out a set of qualities as relevant for the inquiry that is currently carried out. In so doing, existential propositions state the significance of the objects that are the referents of the terms: they establish the facts of the matter.

I have rephrased here Dewey's views in a more contemporary vocabulary, saying that the ultimate function of existential propositions is to provide empirical evidence for theoretical hypothesis. Speaking of evidence – as Dewey himself sometimes does – is helpful in shedding light on the complexity of that notion. As is well known, Dewey articulates an unitary account of inquiry: he maintains that the *structure* of inquiry is the same both in common sense and in science. Among other things, this assumption is central to Dewey's program of extending the scientific approach to social, political and moral issue – an extension which is possible if and only if there is no substantial cleavage between the two forms of inquiry.

Now, one of the drawbacks of Dewey insistence on the continuity between common-sense and scientific inquiries is that the highly refined nature of evidence risks to pass unnoticed. Since common-sense inquiries are usually quite simple – as a consequence of their reliance on common-sense concepts – the process of construction and assessment of evidence is more often than not rudimentary. Clearly, even in the most unsophisticated cases, the significance of evidence for the inquiry is to be evaluated. Nonetheless, such evaluation is mostly habitual and conducted for practical and shortsighted purposes.

The notion of empirical evidence (with its close connection with scientific theorizing) serves, therefore, as a reminder of the fact that the content of existential propositions is entirely constructed within the process of inquiry. This insight – which is now held as platitudinous, but was not that clear at the time in which Dewey wrote his *Logic* – is grounded in the assumption that empirical evidence does not imply or consist in givenness. What is given, properly speaking, is the whole situation, while evidence is constructed within the situation, for the purpose of dealing with the problem at stake. Dewey is explicit on this point. "That which is 'given' in the strict sense of the word 'given'," he remarks, "is the total field or situation". "The given in the sense of the singular, whether object or quality", on the contrary, "is the special aspect, phase or constituent of the existentially present situation that is selected to locate and identify its problematic features with reference to the inquiry then and there to be executed". In this sense, he concludes, evidence is "*taken* rather than given" (LW12, 126).

In Dewey's terminology, evidence is a logical notion. This means that the function of being evidence accrues on some raw material as a consequence of its being employed as a tool in inquiry. So, a consistent and complete theory of evidence should be able to account for the process through which a certain element becomes capable of performing that logical function. In particular, it has to answer two different yet interrelated questions: what is the raw material out of which evidence is constructed; and what are the functions that evidence performs within inquiry. The first question is 'metaphysical', in Dewey's sense of the word; the second belongs to philosophy of science narrowly conceived.

Let's start with the first question. Things have qualities, and – according to Dewey's metaphysics – qualities are particulars: as Dewey writes, "[i]mmediate qualities in their immediacy are unique, non-recurrent" (LW12, 248). Taken by themselves, qualities do not have, therefore, any logical import at all: they are what they are, full stop. However, though they are unique and unrepeatable, qualities can be used to reliably discriminate objects and events as they have the capacity to act as signs. It is a *fact* that there are regularities in the patterns of qualities. The existence of these patterns warrants the inference from a set of qualities that is currently experienced to a different set of qualities that are not existentially present. As Dewey notes, "the presence of certain immediate qualities is so conjoined with certain other non-immediate qualities that the latter may be inferred" (LW12, 250). It is this fundamental – and, in the last analysis, unaccountable – feature of our world that makes both signification and evidence possible.

The issue to be dealt with is, therefore, to understand the proper nature of the qualities on which the logical property of being an evidence accrues. Now, the problem with Dewey's metaphysics is that it is sketchy, bizarre, and rather difficult to locate. At first glance, it seems to have some points of contact with contemporary Humean Supervenience thesis about the fundamentality of qualities. In Lewis's now classical formulation, Humean supervenience is defined as:

> the doctrine that all there is to the world is a vast mosaic of local matters of particular fact, just one little thing and then another. (. . .) We have geometry: a system of external relations of spatio-temporal distance between points. Maybe points of spacetime itself, maybe point-sized bits of matter or aether or fields, maybe both. And at those points we have local qualities: perfectly natural intrinsic properties which need nothing bigger than a point at which to be instantiated. For short: we have an arrangement of qualities. And that is all. There is no difference without difference in the arrangement of qualities. All else supervenes on that (Lewis 1986, ix–x).

By reading the first three chapters of *Experience and Nature*, one may be led to argue that Dewey would agree with contemporary Humeans on the idea that "all there is to the world is a vast mosaic of local matters of particular fact", with no room for genuine modal properties. In that text, indeed, Dewey states that the world is composed of spatio-temporal qualities. And, similarly to Lewis, Dewey believes that the world is a vast mosaic of local, particular qualities.

Contrary to Lewis, however, Dewey does not believe that qualities should be conceived of as spacetime *points*. Dewey's metaphysics is centered on the notion of event, and what there is is a plurality of irreducible events.[10] "Every existence is an event," Dewey remarks, and the perdurance in time is its distinctive feature (LW1, 63). Now, every event is unique and unrepeatable in its specific individuality: "in every event," Dewey writes, "there is something obdurate, self-sufficient, wholly immediate, neither a relation nor an element in a relational whole, but terminal and exclusive" (LW1, 74). It is only because all the events "have something qualitatively their own" that they can be distinguished from logical abstractions (LW1, 75). Immediate qualities are the mark of reality.

From this thesis Dewey draws the conclusion that things are made of "irreducible, infinitely plural, undefinable and indescribable qualities" (LW1, 74). These qualities are the basis of all possible connections and of every possible conceptual discrimination or distinction. This point is highlighted by the distinction that Dewey draws between quality and order: "[t]emporal quality," he remarks, "is however not to be confused with temporal order", since "[q]uality is a quality, direct, immediate and undefinable," while "[o]rder is a matter of relation, of definition, dating, placing and describing" and is "discovered in reflection, not directly had and denoted as is temporal quality" (LW1, 92).

In this sense, the connections that can be discovered among qualities are to be read as mere regularities. It follows therefore – to go back to the point – that the traits or characteristics that are used to make inferences do not have any kind of strong

[10]It should also be remarked that there seems to be no footing in Dewey's philosophy for introducing the distinction between sparse and abundant properties. According to Dewey, indeed, any quality is final, independently of its level of complexity (LW1, 82). Consider the notion of tertiary quality. This is a point in which any direct comparison between Dewey and contemporary metaphysical approaches simply cannot be made.

metaphysical grounding. Indeed, if the fundamental level of reality is made up of irreducible and indescribable qualities, their relations cannot be nomically necessary since there is nothing in the fundamental qualities that ground those connections. Consequently, the property of having evidential value is purely logical, in the sense that it does not need any metaphysical warrant to hold.

There are some elements in support of that interpretation. On a general level, this conclusion is clearly compliant with the idea of the autonomy of inquiry. But other arguments can be put forth. Take, for instance, Dewey's account of causation as formulated in *Experience and Nature*. According to what Dewey writes in that text, he considers the concept of causation as originating from labor and the use of tools. The fact that an element can be used to bring about some effect is said to be a "more adequate [ground] for acceptance of belief in causality than are the regular sequences of nature or than a category of reason" (LW1, 73). It is difficult to cash out the import of Dewey's statement since many different readings are possible: however, it seems safe to conclude that Dewey excludes the principle of causality to be grounded in the metaphysical structure of reality. Such interpretation is corroborated by what Dewey says in a later passage of *Experience and Nature*, when it is stated that "if Hume's lesson has been learned as well as we flatter ourselves it is learned, we should be aware that any matter of causation refers to something extrinsic, to be reached by inquiry and inference" (LW1, 254).

At the same time, however, Dewey seems committed to a radically different metaphysical view, i.e. to some sort of metaphysics of powers. The key notion is that of disposition. Now, as contemporary debates on the topic have shown, the metaphysics of dispositions can take two different forms. One version holds that a property can be categorical and dispositional in one: each natural property is, therefore, a power and a quality at the very same time, and the distinction between the categorical and the dispositional is a difference in the ways in which a property is described.[11] Call this view the powerful qualities view.[12] The other version, on the contrary, identifies properties with dispositions, and concludes that properties are powers. Call this view the pure powers view.[13]

It must be admitted that Dewey's philosophy does not have the conceptual resources to state the difference between the two approaches in a sharp way[14]; this said, I think it can be reasonably argued that his position is closer to the powerful qualities view than to the pure powers view. Indeed, because of his emphasis on the metaphysical irreducibility of qualities, it would be difficult for him to hold

[11]See, for instance, Martin (1997). In recent times, however, the 'linguistic' account of the difference between categorical and dispositional has been called into question on the ground that it would make such distinction *response-dependent* (Molnar 2003, 155–156).

[12]See, among the others, Martin (1997), Martin and Heil (1999), Heil (2003, 2005), Jacobs (2011), and Ingthorsson (2013).

[13]See Shoemaker (1980), Hawthorne (2001), Mumford (2004), and Bird (2007a,b).

[14]Assuming that such a difference exists. For instance, Taylor (2017) denies that there is a real difference between the two positions.

that properties are powers which consist in dispositions to produce certain effects if properly triggered.

In addition, Dewey explicitly acknowledges that different perspectives can be taken on the same thing. A certain property can be considered as it is in itself – Dewey's technical term here is 'consummation' or 'enjoyment' – or it can be considered as a potential tool for producing a certain effect. Both attitudes are legitimate since both grasp a genuine aspect of reality. In Dewey's terminology, the first attitude is aesthetic, and is concerned with the immediacy of qualities; the second one is cognitive, and deals with its manifold connections. Both of them are ways in which a certain property manifests itself.

According to this reading, the logical property of being an evidence – or better said, of being an evidential *trait* or *characteristic*[15] – accrues on an immediate quality because the latter is part of a complex property, which consists also of a disposition to produce other qualities as its effects. The following quotations exhibit this view:

> 1) Crude materials must possess qualities such as permit and promote the performance of the specific operations which result in formed-matter as means to end. But (1) these qualities are but potentialities, and (2) they are discovered to be the potentialities which they are only by means of operations executed upon them with a view to their transformation into means-to-consequences (LW12, 385–386).

> 2) the qualities observed in consequence of the demonstrative act are just the qualities they are. The nub of any existential identification or characterization of a thing as such-and-such lies in the ground it offers for giving the object a description in terms of what is not then and there observed (LW12, 129).

In this sense, fire has an immediate quality – actually, it has many immediate qualities, but let's make things simple for the sake of clarification – and it also has the disposition to produce smoke if properly triggered in proper conditions. On this ground, since there is a real connection between the immediate quality of fire and the latter's power to bring about smoke, the former can be taken as reliable evidence of the smoke. At the same time, precisely because of the nature of that connection, smoke can be taken as reliable evidence of fire. The relation is reflexive.

Now, I have said above that Dewey's metaphysics is too sketchy to be easily classified and located. So, I am well aware that the interpretation that I am going to

[15]A terminological clarification is in order. Dewey defines the notions of trait or characteristic, quality, and property in the following terms: "We are thus enabled to make definite the logical differences between quality, characteristic trait, and property which have previously been noted. 'Turning paper red', is, as the object of a particular observation, a quality. As enabling reasonably safe inference to be made as to the occurrence of other qualities under certain conditions, it is a distinguishing trait or characteristic descriptive of a kind. It becomes a property when it is determined by negative as well as positive instances to be a constant dependable sign of other conjoined characteristics. It then belongs inherently to all cases of the kind" (LW12, 291–292). *Characteristic*, or *trait*, and *property* are to be understood – at least, according to the reading I defend in these pages – as two different manifestations, at two different levels of complexity, of the same *quality*. Please compare this definition to what Dewey says about subject-matter taking different forms (subject-matter, content and object) in different phases of inquiry.

advance is severely underdetermined by the textual evidence available. Nonetheless, I am convinced that the best way to characterize Dewey's metaphysical views is to see them as converging toward a form of dispositional realism. Dewey's logical notion of potentiality, together with his metaphysics of events, points to the view that the raw materials out of which evidence is constructed are powerful qualities, and that the evidential capacity of a trait is grounded in some real connection that holds between a quality and the manifestation of the disposition associated to that quality, which occurs when the disposition is adequately triggered. But how is it possible? Evidence is a logical notion; that means that evidence originates and performs its function within inquiry. So, in which sense can Dewey's dispositionalism account for the process of construction of evidence?

It might be argued that this reading is flawed, for two reasons. On a general level, it seems to go against the grain of Dewey's attempt to sever metaphysics from logic, that is, to deny that inquiry needs any metaphysical backup. More specifically, it seems to fall short of providing a satisfactory and fine-grained explanation of the process of construction of evidence. Take, for instance, the property of being homeothermic. What is the powerful quality which grounds the capacity of the property 'being homeothermic' to enable the identification of a certain animal as a bird, thus transforming that quality into a *trait* or *characteristic* – a logical element – which is distinctive of a particular kind? Or, to give another example, take the property 'having wings', which, as Dewey remarks, was once considered as reliable evidence of the property 'being a bird', to such an extent that one was considered entitled to draw the inference 'since that animal has wings, it is a bird'. Again, what is the disposition connected to the quality of having wings that produces the logical and evidential trait of "having wings", which grounds the inference from the latter to the property "being a bird"? This line of reasoning strikes us as absurd.

An answer to those questions cannot but be tentative and highly speculative. As I read it, Dewey's solution to this problem is that of adopting a broader conception of powers, according to which a disposition can be triggered in many different ways, and produce many different effects as a consequence of its entering into different contexts and participating in different courses of action. A conception of this kind is at odds with some standard approaches: in the contemporary debate on the nature of dispositions, it is sometimes held that a disposition is individuated by the pair composed of its stimulus condition S and its specific manifestation M – be it actual or not. So, for instance, for a glass to be fragile it has to shatter when struck; similarly, for a substance to be soluble, it has to dissolve when put into water. The disposition of fragility is thus identified by its stimulus condition S (being struck) and its manifestation M (to shatter); the disposition of solubility is identified by its stimulus condition S (being put into water) and its manifestation M (to dissolve); and so on.

In any case, there is no universal consensus about the possibility to treat every disposition in this way.[16] Indeed, it seems highly plausible that some dispositions –

[16]See, in this regard, Williams (2010) and Vetter (2013).

the so-called multi-track dispositions – cannot be identified by a single pair of stimulus and manifestation, and that a plurality of stimulus conditions and manifestations is to be introduced in order to account for them. In contemporary philosophy of science, Cartwright's notion of capacity is probably the most authoritative instance of this kind of approach. In her book *The Dappled World: A Study of the Boundaries of Science*, she explicitly acknowledges that while disposition terms are usually conceived as "tied one-to-one to law-like regularities", her notion of capacities has to be understood in a more liberal way:

> [C]apacities, as I use the term, are not restricted to any single kind of manifestation. Objects with a given capacity can behave very differently in different circumstances. Consider Coulomb's law, $F = -q_1q_2/4\pi\epsilon_0 r^2$, for two particles of charge q and q2 separated by a distance *r*. [...] For here let us just consider what Coulomb's law tells us about the motions of the particle pair. It tells us absolutely nothing. Before any motion at all is fixed, the particles must be placed in a special kind of environment; just the kind of environment that I have described as a nomological machine. Without a specific environment, no motion at all is determined (Cartwright 1999, 59).

Leaving aside the reference to Cartwright's technical notion of nomological machine, which is not particularly relevant for the issue here at stake, the point that I would like to stress is that, according to Cartwright's approach, a capacity is not identified by a *single* kind of manifestation. Rather the contrary, depending on the different stimuli that trigger the capacity, the latter can manifest itself in a plurality of ways. As I read it, Dewey's account of dispositions shares this fundamental insight, and extends it in a new and original direction.

More precisely, I read Dewey as suggesting considering existential propositions – in which the evidence of the case is formulated – as one of the possible manifestations of a powerful quality. Take, for instance, the most classical example of capacity or disposition, the fragility of a glass. According to the most restrictive account, a disposition is triggered by its specific stimulus, and it is individuated by its specific manifestation. So, that disposition is triggered by the event of being struck, while shattering is its specific manifestation. According to the multi-track account, the disposition can be triggered in many different ways – by being struck, but also by being sung to by a soprano – and it can manifest itself in different manners. According to Dewey's extended multi-track account, the disposition can be triggered in many different ways and it can manifest itself in different manners; but one of its manifestations is now to appear as a predicate in an existential proposition in which a potentiality is provisionally attributed to a certain element. The powerful quality "fragile" can be triggered by a certain stimulus – in the case of inquiry, the unique problematicity of the situation which prompts the inquirer to adopt a reflective stance towards the objects of its environment – and manifests itself in the construction of evidence which is expressed by an existential proposition.

I do not want to push the analysis too far given the limited textual evidence available. I limit myself to emphasizing three points that are particularly relevant for my purposes. First of all, it should be clear that the idea that a disposition can manifest itself at a linguistic level is made possible by the fact that, according to Dewey, language is, first and foremost, linguistic *behavior*, namely a particular kind of behavior through which the organism transacts with its environment. In this sense, as a mode of activity, language stands on the very same level as the physical actions of hitting or throwing a certain object. In this sense, my reconstruction of what I have called Dewey's extended multi-track account of dispositions is in continuity with his overall naturalism – which counts as a confirmation (though only indirect and partial) of the plausibility of the former.

Secondly, it is worth noting that such an account has a deep metaphilosophical value. Indeed, stressing that a certain quality has the disposition to manifest itself in, and as, an existential proposition which states the evidence of the case is one of the ways in which Dewey rejects representationalism. It means that the relation between a certain feature of the world and the proposition which formulates it linguistically is not to be understood in terms of correspondence, but rather in terms of expression or articulation. Clearly, this does not mean that – being a metaphilosophical stance – it purports to provide some criterion to distinguish between those powerful qualities that can be articulated in a reliable evidence and those that cannot.[17] Any criterion of this sort has to be purely empirical, and cannot help but being intrinsically contextual.

Finally, as the previous examples show, the kind of extension that Dewey needs in order to make his multi-track account effective for the purpose of explaining linguistic evidence has remarkable philosophical consequences. Indeed, it implies that any quality that can be used as evidence in a course of inquiry is a powerful quality. This is made clear by the following statement:

> An actual, immediate quality thus becomes a sign of other qualities that will (or would) be actual *if* additional operations producing conditions for new modes of interaction were performed. When it is said, for example, "This is iron", the significance of the qualification *iron* consists of potentialities not then and there actualized. The qualities of "this" are actual. But they are taken not in their bare actuality but as evidential signs of consequences that will be actualized when further interactions are instituted (LW12, 252).[18]

---

[17]It is precisely because not every quality is reliable evidence of every other property that some principles of classification are sounder than others. So, for instance, the classification "all winged creatures are avian" takes the trait "having wings" as reliable evidence for inferring "being a bird". That connection was once held as reliable, even though it is now known that is wrong – bats have wings, but they are mammalians, to name only one remarkable example. A different sets of traits have to be singled out as definitory of the kind "bird". This is a point which Dewey puts great emphasis on. See, for instance, (LW12, 294–295).

[18]See also the following important passage, in which the relations between qualities and dispositions on a logical plane are further articulated and clarified: "*The determination that a singular is an enduring object is all one with the determination that it is one of a kind.* The identification of a sudden light as a flash of lightning, of a noise as the banging of a door, is not grounded upon existential qualities which immediately present themselves, but upon the qualities with respect to the *evidential function* or use in inquiry they subserve. What is recurrent, uniform, 'common', is the power of immediate qualities to be signs. Immediate qualities in their immediacy are, as we have seen, unique, non-recurrent. But in spite of their existential uniqueness, they are capable, *in the continuum of inquiry*, of becoming distinguishing characteristics which mark off (circumscribe) and identify a *kind* of objects or events. As far as qualities are identical in their functional force, as means of identification and demarcation of kinds, objects are of the same kind no matter how unlike their immediate qualities" (LW12, 248–249).

The distinction between categorical and dispositional properties is, therefore, called into question; and the notion of disposition is so modified and stretched as to be functional to the formulation of a linguistic account of evidence. A certain quality, which is unique and unrepeatable, has the capacity – as a consequence of its existential connections – to manifest itself as a *characteristic* or *trait* which is descriptive of a particular kind.

This is Dewey's account of the process of construction of evidence from what he calls 'raw material', that is, the concrete and unrepeatable qualities of things – or, alternatively put, of the process of transformation of existential features of reality into logical entities. Let's now turn to the other question that Dewey's theory of evidence must address, namely, what are the functions that evidence plays in inquiry.

As has been repeatedly remarked, it is in the context of inquiry that unique and unrepeatable qualities become characteristics or traits. The function of a set of characteristics is to "mark off (circumscribe) and identify a *kind* of objects or events": a certain quality is discovered to be conjoined with other qualities that will show up or be produced if some operations are performed; then, it is discovered that the connection is sound, repeatable and counterfactually solid; finally, the identification of the kind to which the thing or event possessing that quality belongs is established (LW12, 249). Accordingly, it is the act of identification that entails the constitution of a quality as an evidential datum: qualities acquire evidential value when their immediacy is overlooked, and their 'significance stability' is taken into account. This means, among other things, that it is the capacity of a quality to bear a functional weight rather than its distinctive features as a quality that makes it possible for some evidential function to accrue on it.

Now, though rather simple, this approach is fruitful: it leads to a conception of evidence that is particularly interesting in light of contemporary debates since it solves – or dissolves – some problems that have troubled traditional accounts. In the following pages I will focus on two of those aspects, and, for the sake of stressing their theoretical relevance, I will present them as possible solutions to contemporary problems. In so doing, I will also shed light on the functions that Dewey attributes to evidence in the context of inquiry, and assess the consistency of his proposal.

Let's start with the first issue. As Julien Reiss has pointed out, two different conceptions of evidence are commonly referred to in contemporary debate. On the one hand, evidence may be understood as *supporting* a certain hypothesis, where the hypothesis is not for this reason accepted or endorsed. In this sense, evidence works as a mark or sign of the hypothesis being true, so that $P(H|E) \geq P(H)$. Nonetheless, the value of P(H|E) does not need to be equal to 1. The second conception is probably closer to the naive way of understanding evidence: in this case, evidence is taken as a *proof* of a certain hypothesis; it warrants the belief in the hypothesis. Accordingly, it holds both that $P(H|E) \geq P(H)$ and that P(H|E) = 1.

Reiss argues – and I think it is right on this point – that the fundamental requirement of any consistent theory of evidence is that it succeeds in accounting for both the conceptions of evidence, thus providing the conceptual resources to distinguish them in a clear and satisfactory way (Reiss 2015, 342–343). I think that Dewey's notion of existential proposition is capable of accomplishing that goal.[19]

In particular, Dewey's philosophy of science is rich enough to account for both the conceptions of evidence. Clearly, the explanation of their difference is functional. As has been remarked above, it is of the nature of any proposition – either universal or existential – to be affirmed but not asserted. Up to now, I have not distinguished among the different degrees of strength of affirmation and, consequently, of evidence. However, because of the temporal nature of the process of problem-solving, it is clear that the strength of affirmation changes throughout inquiry. While in a earlier stage the function of evidence is that of supporting a certain hypothesis – whose function is to bring about further evidence – in a later stage evidence becomes more and more logically compelling.[20] When the evidence thus gathered reaches a satisfactory level of coherence, it warrants the hypothesis. In so doing, the *content* of existential and universal propositions turns into the *object* of a judgment, where inquiry ends.

The idea of a functional correlation between evidence and hypothesis implies that evidence is evidence *for* a hypothesis. Dewey is explicit that something counts as evidence only insofar as, and to the extent that, it can pave the way for the formulation of an explanatory hypothesis. If it does not perform this function, it does not have logical import, properly speaking. It is worth noting, however, that the idea of a functional continuity does not rule out the possibility of alternative hypotheses. The possibility of alternative hypotheses has been signaled as a problem for some accounts of evidence. The problem can be formulated as follows:

> whatever rule is specified for positively appraising H, there will always be rival hypotheses that satisfy the rule equally well. Evidence in accordance with hypothesis H cannot really count in favor of H, it is objected, if it counts equally well for any number of (perhaps infinitely many) other hypotheses that would also accord with H (Mayo 1996, 174).

Now, I believe that this problem does not pose any serious trouble to Dewey's approach. Leaving aside the most radical consequences that can be drawn from it, which lead to a form of radical underdetermination, Dewey recognizes the existence of alternative hypotheses as an essential feature of inquiry. "Whenever

[19]More precisely, in the article *A Pragmatist Theory of Evidence*, Reiss lists four requirements that a good theory of evidence should satisfy: (1) it distinguishes support and warrant; (2) it provides an account of evidential support; (3) it provides an account of warrant that allows warrant to come in degrees; (4) it applies to nonideal circumstances typical of science in practice. Now, my contention is that Dewey's pragmatist approach is rich enough to satisfy all these requirements (Reiss 2015, 344).

[20]Obviously, 'logically' is used in Deweyan sense. It does not mean that the hypothesis is a logical consequence of the evidence, so that $P(H|E) = 1$; rather, it means that the evidence collected throughout inquiry is capable, to quote Dewey's own words, of "form[ing] an ordered whole in response to operations prescribed by the ideas they occasion and support" (LW12, 117).

there is genuine deliberation," he remarks, "there are alternatives at almost every step of the way" (LW12, 165). Again, the key notion here is that of temporality of inquiry. Inquiry originates from an indeterminate situation. Now, at this starting level, the gathered evidence points towards a certain possible hypothesis. However, being indeterminate or problematic means that different courses of action can be envisioned, which are equally plausible in light of the data available. The elaboration and choice of a hypothesis is, therefore, provisional: the affirmation of the hypothesis A – say, 'Z is the cause of Y', whose logical import is something like "Z might be the cause of Y" – is made only tentatively, and immediately suggests some extremely general experimental operations aiming at detecting whether or not that hypothesis is sound. This is due to the fact that, according to Dewey, a hypothesis is nothing but a possible general mode of action. Accordingly, the possibility of alternative hypotheses means the possibility of alternative ways of handling the problematic situation.

If the hypothesis A is partially confirmed by the evidence that it produces by directing a course of experiment, then its affirmation gains logical force at detriment of other, alternative hypotheses.[21] At the same time, however, provided that the problem under investigation is not so simple as to be solved at the first attempt, the evidence will require modification and further refinement of the hypothesis. Such modification may be more or less radical, but its cumulative character entails that something is preserved throughout the process. What is preserved is precisely the evidential character of data. So, for instance, if the hypothesis 'Z is the cause of Y' is partially confirmed by evidence, some kind of connection between Z and Y, if ascertained, becomes a fact that puts constraints on the successive steps of elaboration of the hypothesis.

But why is this particular hypothesis A to be affirmed and, therefore, provisionally accepted? This is the final question that a pragmatist theory of evidence must

[21]Something more can be said in this regard. For a proposition to be grounded, Dewey writes, "alternative possibilities must be ruled out" (LW12, 189). The point that Dewey is trying to make – which is rather classical, as a matter of fact – is that affirmation cannot be disjointed from negation: when something is affirmed, all the other elements of the logical space are indirectly denied. Take the existential proposition "this is red". Dewey writes: "[v]alid determination that "*this* is *red*" depends upon (1) exhaustive disjunction of alternative possibilities of color and (2) upon elimination of all other possibilities than the one affirmed, the elimination resulting (3) from a series of hypothetical propositions such as "if blue, then such and such consequences," etc., in contrast with the proposition "If red, then such and such other and differential consequences" (LW12, 189). Accordingly, the true logical form of the hypothetical proposition that grounds the conclusion "This is red" is something of this kind: "*Only* if this is red, will observed phenomena be what they are" (LW12, 189). The logical force of that 'only' derives from a series of eliminations of the alternative hypotheses which are expressed in and through negative propositions. The kind of elimination that is thus performed is related to a series of experimental operations whose function is to provide new and better evidence. This task can be accomplished only because alternative hypotheses play a role in the process of determination and elaboration of the selected hypothesis.

address in order to provide a satisfactory and complete account of the function of evidence in inquiry, which is capable of being applied to the nonideal circumstances of scientific practice.

Usually, this issue is framed in terms of the assessment of prior probabilities. According to the standard subjectivist Bayesian approach, an agent can assign any value between 0 and 1 to the hypothesis. That value expresses the likelihood of the events stated by the proposition. This is not, and cannot be, Dewey's approach. The subjectivist conception of probability – as Savage puts it – "is a code of consistency of the person applying it, not a system of predictions about the world around him" (Savage 1972, 59). Now, because of his assumption of the logical primacy of situation over belief, Dewey cannot follow this line of reasoning. To insist that the attribution of priors is a personal and subjective matter would entail that the first step of inquiry has no logical import, which is a conclusion that Dewey could hardly accept.

Dewey's answer to this problem is, therefore, a direct consequence of the assumption that evidence is not taken for granted, but is constructed for the purpose of formulating a hypothesis that could lead to new and more reliable evidence. According to the temporal pattern of inquiry, the choice of an hypothesis – or, differently said, the assignment of priors to alternative hypotheses – is grounded in the kind of evidence that is available at the initial phase of inquiry. Dewey is explicit on this point:

> The first step in answering this question [namely, how is the formation of a problem so controlled that further inquiries will move toward a satisfactory solution?] is to recognize that no situation which is *completely* indeterminate can possibly be converted into a problem having definite constituents. The first step then is to search out the *constituents* of a given situation which, as constituents, are settled. Since they are settled or determinate in *existence*, the first step in institution of a problem is to settle them in *observation*. There are other factors which, while they are not as temporally and spatially fixed, are yet observable constituents [...]. All of these observed conditions taken together constitute "the facts of the case." They constitute the terms of the problem, because they are conditions that must be reckoned with or taken account of in any relevant solution that is proposed (LW12, 112).

In other words, the context of inquiry puts normative constraints on the prior probabilities that can be assigned to the various hypotheses, these constraints being established by those features of the indeterminate situation that constitute the unique and unrepeatable indeterminacy of that specific inquiry. Obviously, the contents of this initial observation count as evidence only insofar as, and to the extent that, they prove themselves to be effective in directing inquiry towards a possible rational solution of the problematic situation – they are intrinsically revisable.

In this sense, Dewey's theory of evidence also satisfies a further requirement imposed by Mayo, namely, that "an adequate account of experimental testing must not begin at the point where data and hypotheses are given, but rather must explicitly incorporate the intermediate theories of data, instruments, and experiment that are required to obtain experimental evidence in the first place" (Mayo 1996, 212).

4.4 The Constitutive Function of the A Priori

In the previous section, while discussing the process of construction of evidence, I have remarked that, within the context of inquiry, an existential quality becomes an evidential trait or characteristic which enables the identification of the object as one of a kind. The question that I would like to address in this section concerns the ground of such an act of identification.

In order to adequately introduce this issue, let's consider the example made by David Stump in the opening pages of his book *Conceptual Change and the Philosophy of Science*. He asks the reader to imagine a researcher working in a biology lab who opens a new box of Petri dishes containing agar prepared for growing cultures. The researcher finds that one of the Petri dishes contains a growth; so she throws it away without hesitation. Clearly, the growth is taken as evidence of the fact that the dish was contaminated, and that some microbes had entered in a medium that was supposed to be sterile.

Stump presents such a reconstruction as uncontroversial. But why should it be so? Why, for instance, does the researcher pay no heed to the hypothesis of that growth as evidence of spontaneous generation in a sterile medium? In so doing, Stump remarks, "[t]he researcher has, by implication, rejected the second explanation without thinking about it at all, when throwing the Petri dish in the garbage, thereby foregoing a chance for a Nobel Prize" (Stump 2015, 1). So, there must be something that warrants her conclusion that the Petri dish is contaminated, and justifies her behavior.

One possible solution to that problem is that the researcher takes the growth as evidence of some sort of contamination simply because (a) there is a real connection between the first and the second event; (b) she knows that such a connection exists; and (c) she believes that that connection, which she has verified in the past, will hold in the future. A reply along this line is obviously correct, but it is of no help in this case. Indeed, the point at stake is neither to ascertain whether the connection exists nor to justify the belief in that connection; rather, the point is that of understanding the logical ground – i.e., the ground *within* inquiry – which enables the inquirer to use the growth as evidence of contamination, and to identify that Petri dish as one of the kind "contaminated dish". In other words, the point at stake is not that of providing the justification of induction, but that of singling out those elements that perform a constitutive function within inquiry, thus providing the criteria for transforming qualities into evidential signs. The two issues – justifying induction and understanding the constitutive element on inquiry – are clearly related, as Dewey openly acknowledges, but should not be conflated.[22]

[22] I will not deal with Dewey's account of induction here. A satisfactory analysis of this issue can be found in Burke (2002). As is quite well-known, Dewey's theory of induction has been strongly criticized by Brodbeck in the article Brodbeck (1949). For a reply to Brodbeck, see Mayeroff (1950). I hope to address this topic in a future work.

Dewey becomes aware of the relevance of the constitutive element in inquiry relatively lately. It is only from the mid-1930s' – and, in particular, with the publication of the three articles *Characteristics and Characters: Kinds and Classes*, *What Are Universals?*, and *General Propositions, Kinds, and Classes* in 1936 – that he realizes the importance of making room for some sort of constitutive principle of scientific knowledge. The most refined result of that process of revision of some of the fundamental assumptions of his logical theory can be found – not surprisingly – in the *Logic*, where Dewey articulates an account of what goes under the name of operational a priori. In so doing, Dewey enters into the debate on the possibility of preserving a role for a priori within a naturalistic and historicist framework.

The original formulation of this insight is due to Hans Reichenbach who, in his influential book *The Theory of Relativity and A Priori Knowledge* (1920), distinguishes between two different kinds of axioms, namely the *axioms of coordination* and the *axioms of connection*. While the latter are empirical laws that make use of terms whose meaning is well established, the former are not empirical, and their function consists in establishing those meanings. Now, since axioms of coordination constitute the object of knowledge, they are a priori.

However, as Reichenbach is ready to note, a priori can be understood in two different ways. On the one hand, the notion of a priori is essentially connected to the idea of necessary and unrevisable truth. In this sense, what is a priori is said to belong to the very structure of human understanding, and, consequently, cannot be changed, revised, or modified. This way of understanding the a priori is clearly dependent upon a metaphysical conception of human reason, and is therefore no longer tenable in light of the post-Darwinian image of the man in the world. On the other hand, a priori is used as a synonym of constitutive, and refers to the function that certain principles have to make certain other empirical laws possible: a priori means "before knowledge", but that does not imply that what is a priori is "for all time" (Reichenbach 1965, 48). The best example is given by the relation between geometry and physics. Kant believed Euclidean geometry to be true a priori, and he also held that the (synthetic) a priori truth of Euclidean geometry warranted the (synthetic) a priori truth of Newtonian mechanics – or, more precisely, of some of its fundamental principles. After the discovery of non-Euclidean geometry, however, it became clear that Euclidean geometry was not true a priori; nonetheless, in the context of Newtonian physics, Euclidean geometry continues to work as a set of axioms of coordination insofar as it makes possible the confirmation of the empirical laws of that science. Here it is clear that a set of principles can perform a constitutive function without being considered universally and necessarily true. Though conflated by Kant, the two notions are clearly different.

The shift of emphasis from the origin of the a priori principles to their function – which lies at the basis of the idea of a relativized a priori originally advanced by Reichenbach and recently articulated and defended by Michael Friedman[23] –

[23]Much has been written on Reichenbach's reformulation of the Kantian a priori, as well as on the debate that followed the publication of his book. See, among the others, Coffa (2003, Chapter 10),

is strongly consonant with Dewey's views on the nature of inquiry. No surprise, therefore, that Dewey's "rediscovery" of the a priori follows precisely these lines.

Dewey's most explicit recognition of the function of the a priori in inquiry is stated in the opening chapter of the *Logic*, where he remarks that logical forms are postulational. By this notion, Dewey means to refer to the fact that logical forms are stipulated for the sake of conducting inquiry to its successful conclusion. Some sort of a priori principle is active in every human activity: habits of behavior perform the function of anticipating the environmental responses to the actions of the agent. When inquiry reaches a certain level of complexity, as happens in the case of scientific research, one of the preconditions for a successful resolution of the problematic situation is that the habits of reflection of the inquirer are linguistically formulated, made explicit, and, in so doing, explicitly acknowledged as one of the elements of inquiry.[24]

Now, the notion of postulation or stipulation risks to be misleading in that it may convey the idea that Dewey endorses a strong form of conventionalism. As should be clear from what has been said in Chap. 2, Dewey believes that the notion of convention is unproblematic only insofar as it is understood within a practice-centered framework; otherwise it is self-refuting. The point that he wants to make with his insistence on the postulational nature of logical forms is that the choice of certain principles or concepts is conventional in the sense of not being necessitated by any motives external to inquiry; nonetheless, the criteria for assessing their validity cannot be conventional too. In this sense, Dewey remarks, a postulate is "neither arbitrary nor externally *a priori*": it cannot be externally a priory since "it is not imposed upon inquiry from without, but is an acknowledgement of that to which the undertaking of inquiry commits us"; similarly, it cannot be arbitrary "because it issues from the relation of means to the end to be reached" (LW12, 25). Once again, contrary to what the stereotypical image of Deweyan pragmatism may lead one to think, the criteria for assessing the validity of a logical forms do not take into account only the consequences of the decisions made by the inquirer; rather, they essentially refer to the intrinsic normativity of the specific situation in which the inquirer is engaged, as well as to the objective conditions of the situation that inquiry has to address.

Parrini (2002, Chapters 1 and 3), Ryckman (2005, Chapter 2), and the Appendix. The expression "relativized a priori" was coined by Michael Friedman. See Friedman (1999) and, in particular, Friedman (2001) in which the notion of relativized a priori is used to articulate "a modified version of a Kantian philosophy of science", which is capable of acknowledging a place for the idea of constitutive principles of objectivity within a post-Quinean and post-Kuhnian framework (Friedman 2001, 71). Friedman's work has sparked interest in the issue. See, in particular, the remarks made by Mormann in (Mormann 2012), in which the notion of relativized a priori is interpreted in pragmatist terms.

[24]Incidentally, it is interesting to note that, in order to clarify this point, Dewey draws a comparison between logical forms and what he calls the postulates of geometry: "[j]ust as the postulates of [...] geometry are not self-evident first truths that are externally imposed premises but are formulations of the conditions that have to be satisfied in procedures that deal with a certain subject-matter, so with logical forms which hold for *every* inquiry" (LW12, 24).

Logical forms are, thus, operational a priori in that they represent a statement of the conditions to which the inquirer commits herself with the hope of reaching the conclusion of the inquiry. Adopting a postulate means taking the responsibility for acting in a certain way when certain conditions take place. At the same time, for a principle to be empirically or temporally a priori means that it has emerged from previous transactions with the subject-matter, and purports to lead, if used correctly, to the reconstruction of the problematic situation that has called out inquiry.

Clearly, this functional conception of the a priori is remarkably different from the traditional, Kantian understanding of the notion. In more contemporary terms, it can be viewed as a naturalization of the a priori which rejects a great part of Kant's theoretical apparatus. As is well known, in the *Critique of Pure Reason* Kant firstly undertakes the metaphysical deduction of the categories from the table of judgments, and only after such a deduction is completed, he attempts to provide their transcendental deduction. Now, any naturalized account of the a priori cannot help but reject the very idea of a metaphysical deduction; as a consequence of that, however, the relevance of the transcendental deduction is also called into question since the validity of the constitutive element of scientific knowledge can be explained in purely empirical terms, as the refinement of previous, successful inquiries. In this sense, the idea of naturalized a priori can be easily accommodated within a thoroughly naturalistic framework. Stated in other words, there is nothing in the idea of a naturalized and relativized a priori that excludes its assimilation into a Quinean holistic picture of human knowledge as a web of beliefs that, though impinging on experience only at its periphery, is open to revision in any of its components.

Those insights lie at the core of Dewey's theory of the constitutive element of inquiry. Recall the example with which I have opened this section. As said, Dewey's problem is that of understanding how it is possible to use a particular quality as an evidential sign that enables the identification of an object as one of a kind. "A singular as a mere *this*," Dewey writes, "always sets a problem", which is "resolved by ascertaining *what* it is – that is, the kind it is of" (LW12, 249).

The singular quality – the *this* – is unique and unrepeatable. So is the act made by the inquirer in dealing with that quality – for instance, the act of bombarding it with protons or that of putting it under a microscope. However, those operations are *general* ways of acting on the environment: even though they are *exemplified* in singulars acts, they are not themselves singular. As operations, they can recur indefinitely. This means that they are functionally a priori with respect to the present situation. The inquirer has those general modes of behavior at her disposal before engaging in the course of inquiry.

Dewey is clear on this point. The starting assumption of his analysis is the acknowledgment that "qualities become traits descriptive of a kind when they are the consequences of operations which are modes or *ways* of changing and acting" (LW12, 252). The process of construction of evidence out of raw material has been highlighted in the previous section; so I will not deal with it any further. The present formulation simply puts at the forefront of the analysis the structural connections among qualities, general operations or ways of acting, and the consequences of those operations, but it is by no means different from what has been pointed out above. What is interesting to note is the further step that Dewey takes. Commenting that statement, he not only clarifies that there is a difference between the generality which pertains to kinds and the one which is attached to the operations that can be performed by the inquirer; he also acknowledges a relation of dependence of the former on the latter: "the type of generality which constitutes the logical form of the [kind] is derivative," he remarks, "depending upon the generality of the operations executed or possible" (LW12, 253).

Let's focus on this issue. Generality can be conceived of in two different ways. On the one hand, there is the generality of kinds. A set of traits is used to identify a specific group, and to that group some further properties are attributed. Dewey names the propositions which are about kinds or about their characteristics *generic* propositions. On the other hand, there is the generality of the operations. This type of generality is constitutively different from the first one in that it refers to the ways in which the inquirer can act and modify her environment; it has therefore to do with the principles through which the inquirer interprets and pre-constitutes its objects.

It is a commonplace that the same set of qualities can be used as an evidential sign of many different kinds: according to the phlogiston theory, the fact that the combustion soon ceases in an enclosed space is evidence that air has a limited capacity to absorb phlogiston; according to the oxygen theory, the same fact is evidence that it is oxygen which makes combustion possible, since it accepts the electrons released in the oxidation. The difference in the way in which the same set of qualities is constructed into two different evidential signs, as well as of the ontological commitments that follow from the adoption of one or the other theory, depends on the difference in the principles of classification. Dewey names the propositions which are about categories (classes) or about their characters *universal* propositions.[25]

So, within a controlled course of inquiry the unreflective habits of behavior of the inquirer are formulated in universal propositions which set the criteria for identifying the objects under investigation as belonging to a certain kind. The propositions in which such acts of identification are formulated are existential propositions. The latter can take different forms: when a singular object is identified as one of a kind – "*this* is a bird", "*this* is an atom of oxygen", "*this* is an economical crisis" – the proposition is *singular*; its function in inquiry is that of constructing the object as something that has the potentialities established by the particular kind to which the object is said to belong, which will turn into actualities if the object is operated upon in a certain way. However, a kind can be identified as being included in a more extensive one. In so doing, the number of characteristics that can be inferred from the set of qualities that constitutes the smaller kind is greatly enhanced. When a physician formulate the existential proposition "professional runners have hypertrophied hearts", she states a connection between the two kinds, thus committing herself to use those traits that identify a professional runner as

[25] A useful classification of Dewey's propositions is provided in Burke (1994, 176–190).

evidential sign of a hypertrophied heart. In this case, the existential proposition is *generic*. Singular and generic propositions are, therefore, different in the scope of the inferences that they support: contrary to singular propositions, generic propositions enable to organize observed and inferred characteristics in a system.[26] But in both cases, their function in inquiry is the same: they are instrumental to inferring non-observed qualities from a set of observed qualities taken as evidential signs.

Universal propositions work differently. They do not purport to draw any inference about the objects of the situation; as statements of *possible* operations, they do not have existential bearing. However, as Dewey openly recognizes, an universal proposition is not "*merely* a formulation of a way of acting or operating" (LW12, 269). If it were, it would not perform any constitutive function. Rather, its function is that of making it possible for the inquirer to construct the evidence that she needs for the purpose of solving the problematic situation in which she is engaged. To quote Dewey's own words, "[i]t is *such* a formulation as serves to direct the operations by means of which existential material is selectively discriminated and related (ordered) so that it functions as the ground for warranted inferential conclusions" (LW12, 269).

The structure of universal propositions is of the *if-then* form: an universal proposition states a necessary relation between the contents formulated in the apodosis and those expresses in the protasis. Accordingly, the kind of inclusion that is here at stake is not an inclusion between kinds and their observable traits, but rather between meanings (LW11, 105). Dewey is explicit on this: "[b]eing a part of an idea is obviously different from being a part of an existential collection or a member of an existential kind" (LW11, 106).[27] Universal propositions formulate the definitions of the concepts used in inquiry. In this sense, Dewey argues that it is wrong to say that the content of the apodosis follows from that of the protasis; in reality, the if-then proposition provides "the analysis of a conception into its

---

[26] In some passages, Dewey seems to be ready to take a step further, and to acknowledge a sort of expressivist connection between these two types of propositions. Among the other things, this way of framing the issue sheds an interesting light on the relations between science and common sense. Dewey formulates that insight as follows: "Such systematization is one of the chief differences between common sense and scientific kinds. It is this systematic serial relationship which renders the category of *membership* or inclusion applicable to included kinds and not to the case in which a singular is simply identified and demarcated as one of a kind. The proposition of a relation of kinds thus provides the logical ground of the singular proposition. For in the proposition of the form "This is one of a kind," there is *implicitly* postulated that there are other kinds related to the one specified. For characteristics which suffice to ground the reference of *this* to a kind must be such as to demarcate it from other kinds. The adequate grounding of such a proposition demands, accordingly, that related but excluded kinds be determinately established" (LW12, 294).

[27] Dewey considers these two propositions: "Each and every whale, whether observed or not, or whether now existent or not, is a mammal" and "If an animal is cetacean, it is mammalian." Now, he writes, "[w]hen we compare these two propositions as to their logical form, it is evident that the latter expresses a necessary relation of characters and holds whether whales exist or not. The first proposition refers to each and every existence marked by a certain conjunction of traits" (LW12, 256)

integral and exhaustive contents": "the two clauses represent the analysis of a single conception into its complete and exclusive interrelated logical constituents" (LW12, 270).

The function of universal propositions is, therefore, that of formulating and articulating in a linguistic medium the meaning of the concepts used in inquiry. In so doing, rigid constraints are put on the process of construction of evidence and of objectivity, with the latter being the final outcome of the former. Indeed, universal propositions state the conditions that an object *must* meet in order to be an object of a kind. In this regard, Dewey writes:

> a universal hypothetical proposition has the form of a definition in its logical sense. Thus the proposition "If anything is material body, it attracts other material bodies directly as its mass and indirectly as the square of the distance" may read equally well in the linguistic form "*All* material bodies, etc." It is a (partial) definition of being a material body. It expresses a condition which any observed thing must satisfy if the property "*material*" is groundedly applicable to it (LW12, 271).

In light of what has been said, the difference between generic and universal propositions should now be clear. As existential propositions, generic propositions are used to draw inference about some qualities that are directly observed. If the inquirer states the existential proposition "All material bodies attract others material bodies", she is open to take a set of qualities as evidence of the existence of a material body, and then to infer from them that that particular material body has the potentiality to attract other material bodies directly as its mass and indirectly as the square of the distance. It may be, however, that she is wrong: maybe that property does not belong to *all* material bodies, but only to a small subset of them, and consequently that inference is ungrounded. As Dewey notices, history of science is full of examples of this sort.

On the contrary, if the inquirer states the universal proposition "All material bodies attract others material bodies", no matter what an experimental course of activity may reveal about the specific object with which she is dealing, if that object does not attract other material bodies, it is not a material object. Attracting other material bodies is part of the definition of a material body, and consequently that condition cannot fail to be satisfied. This also means that, being rules that state how a concept is used, universal propositions are held immune from disconfirmation. The inquirer cannot be wrong about the meaning of her concepts.

As hinted at above, however, the acknowledgment of the specific function performed by universal propositions within inquiry, as well as the difference in logical form – to use Dewey's terminology – between generic and universal propositions, results in the rejection neither of holism nor of naturalism. Dewey's argument runs as follows. Its starting point is the thesis that universal and generic propositions are part of the same course inquiry, and are functionally related as instruments for the reconstruction of a problematic situation. Accordingly, their difference depends exclusively on the way in which they are employed by the inquirer. Furthermore, in order for an inquiry to be successful, the two functions have to be integrated and cooperate. In this regard, Dewey speaks of a "conjugate relationship" between the two types of propositions (LW12, 278). So, it is true that the operations which constitute the content of the universal propositions determine evidential data by putting constraints on the process of construction of the latter; this is how they express their constitutive function. At the same time, however, "the data which are thus constituted become the tests of operations already executed and the ground upon which new operations (or modifications of old ones) are suggested and executed" (LW12, 273). Consequently, though they are held immune from direct empirical disconfirmation, universal propositions are open to be revised – or even abandoned – once they prove themselves to be ineffective as tools for constituting evidence. I will come back to this point momentarily.

It is clear, therefore, that there is no need to grant universal propositions a privileged epistemological status. The distinction between empirical and constitutive can be formulated exclusively on a purely functional basis. Similarly, no permanent distinction needs to be postulated between constitutive and empirical elements within inquiry. Indeed, in the course of inquiry an element can have its status shifted from empirical to constitutive. Dewey illustrates this point by means of an example drawn from the history of medicine, namely, the construction of the kind *malaria*.

As is well known, the causes of malaria had been known for a long time, as well as some of its possible remedies – long before any scientific explanation of the disease was formulated. The conception "malaria" so constructed had, Dewey remarks, a merely practical value: as a common-sense concept, it merely puts together under one label the different phenomena exhibited in the course of the disease. Its inferential power was, therefore, extremely limited: while the identification of cases of malaria was not problematic since its symptoms are easily recognizable, the inferences based on that notion failed to throw any light on the nature of the disease.

It is only when it was discovered that some diseases are of a parasitic origin that a scientific understanding of malaria became possible. Dewey analyzes at some length the various phases that led to a consistent scientific explanation of the origins and causes of the disease. I will not dwell on the details of his reconstruction since they are not particular relevant for my purposes. What is interesting is, rather, how Dewey accounts for the final step of that process:

> The hypothesis as to inherent connection between the development of the disease and the parasite of the mosquito in the blood was clinched when this empirical fact was experimentally shown to follow from the relation between chemical properties of quinine and the conditions requisite for maintenance of life on the part of the parasite. A universal proposition of the "if-and-only-if, then" was finally grounded as far as any such proposition is capable of conclusive grounding (LW12, 431).

The point that Dewey wants to highlight is twofold. On the one hand, he explicitly acknowledges that, in the course of inquiry, the boundaries between the empirical and the constitutive are constantly revised and defined. The empirically tested connections between the development of malaria and the presence of the parasite of the mosquito in the blood become an universal proposition of the *if-then* form – something of this sort: if the patient presents certain symptoms, then you find the parasite of the mosquito in her blood. Which means that, being formulated in an universal proposition, that relation is taken as a partial definition of the concept "malaria".

On the other hand, however, the case of malaria sheds light on another, more complex function performed by the constitutive element. I refer to Dewey's statement that the hypothesis of an *inherent* connection between malaria and the presence of the parasite of the mosquito in the blood was eventually transformed into an universal proposition when scientists succeeded in proving *experimentally* that that existential connection follows from the relation between the chemical properties of quinine and the conditions needed by the parasite to live and proliferate.

I think that Dewey has something of this sort in mind. It is a fact that the connection between the development of the disease and the presence of the parasite of the mosquito in the blood was empirically ascertained by scientifically trained observation. However, that connection was merely provisional evidence: it purported to provide factual information about the subject-matter under investigation, but its logical value was partially undetermined since it was not clear how much inferential weight that connection could bear. The derivation of that empirical connection from the knowledge of chemistry available at that time dramatically changes the whole scenario. The possibility of experimentally showing that an until-then empirical fact, formulated in a generic proposition, can be derived from an established body of knowledge – which has already reached the status of science, and whose validity is not questioned in the course of the particular inquiry in which the agent is engaged – modifies its logical status. It thus acquires the same logical status of the universal propositions belonging to chemistry from which it is derived.

It is worth noting, in this regard, that Dewey does not speak here of deduction but rather of experimental proof. The grounding of the universal proposition which states the connection between the development of malaria and the presence of the parasite in blood does not result *simply* from its being logically derived from the body of knowledge of chemistry. On the contrary, it issues from an experimental setting in which that connection is tested. In so doing, Dewey not only stresses the primacy and creativity of scientific practice; he also remarks that the application of a certain set of already established universal propositions to a new case – not immediately translatable into the language of the theory – necessarily entails some aspect of originality.[28]

[28]The last remarks highlight a feature of Dewey's account of the constitutive element of science which is of some relevance for contemporary philosophy of science. Recall the example made by Stump, which I have quoted at the beginning of this section. Commenting on the philosophical import of that example, Stump remarks that the latter can be read in two different ways, which points to two different and alternative philosophical frameworks. I quote the passage in its entirety: "[t]here are at least two ways of looking at the setting up of criteria in this manner. The philosopher Arthur Pap discusses this in terms of meaning, so that he would want to say that the definition of an uncontaminated Petri dish includes the idea that there are no microbes in the dish before they are placed there in the course of research. We could, however, leave out any discussion of the meaning of terms and simply say that the presence of microbes in a new Petri dish has become a criterion of the contamination of the dish. The second option focusses on the practice of science, rather than semantics, which provides a picture of science that is more historically accurate" (Stump 2015, 2).

In recent debates over the notion of relativized a priori, one of the functions that are usually attributed to the constitutive element of knowledge is that of making it possible to formulate empirical laws. This point is explicitly made by Friedman in his *Dynamics of Reason*, where he states that "the role of what I am calling constitutively a priori principles is to provide the necessary framework within which the testing of properly empirical laws is then possible" (Friedman 1999, 83). With the malaria example, however, Dewey seems to take a different road. In a sense, it is true that he admits that chemistry provides a framework within which the testing of properly empirical laws is possible. Nonetheless, those principles are not constitutive in the sense of being necessary in order to formulate empirical observations: rather, they are constitutive in the sense of making it possible to transform those empirical observations, formulated in existential propositions, into scientific principles which state the meaning of the terms employed in inquiry.

This is a form of constitution which is different from the one described by Friedman. Indeed, Dewey is explicit that the discovery of empirical connections is not dependent on the adoption of chemistry as a framework of analysis; the functional autonomy of existential propositions excludes any strong transcendental approach. Constitution is here understood as grounding: the function of the constitutive element is to ground an existential proposition, thus extending the ways in which that content – the connection between the development of the disease and the presence of the parasite of the mosquito in the blood – can be used in inquiry.

Before moving on to the final section of this chapter, in which Dewey's conception of objectivity is addressed, I would like to call attention to one further point that I have left untouched. As should be evident from what has been said heretofore, Dewey's theory of functional a priori has many points of contact with Lewis's notion of pragmatic a priori. Both Lewis and Dewey believe that the constitutive element provides patterns of interpretation that have to be "determined in advance of the particular experience to which they apply in order that what is given may have meaning" (Lewis 1929, 230); they share the assumption that since "a priori is something made by mind, mind may also alter it", provided that not too much epistemological or metaphysical load is put on the notion of mind (Lewis 1929, 233); they also agree on taking concept as the proper unit of analysis of the constitutive element of knowledge. As Lewis remarks, "[c]oncepts thus represent what mind brings to experience" – and it is not by chance that his position is usually labeled "conceptual pragmatism" (Lewis 1929, 230). More fundamentally, they both accept the basic pragmatist insight that knowing is a mode of action, to such an extent that the following sentence of Lewis could have been written by Dewey: "[a]t the bottom of all science and all knowledge are categories and definitive concepts

What is relevant to note is that, according to Dewey's approach, these two ways of interpreting the Petri example are not alternative. Rather the contrary, the semantic reading clarifies the function that the constitutive element of knowledge plays in the concrete practice of science.

which represent fundamental habits of thought and deep-lying attitudes which the human mind has taken in the light of its total experience" (Lewis 1923, 176).[29]

However, all those similarities notwithstanding, there are some relevant points of disagreement between the two approaches that prevent from considering Dewey's account of the constitutive element as a mere restatement of Lewis's views. Some of their differences are terminological, and trace back to Dewey's rejection of the epistemological language that Lewis adopts. As hinted at above, Dewey would have not formulated his position in terms of the relation between the mind and the given. Since I have dealt with this topic extensively in the previous chapters, I won't dwell on it any further here. Suffice to say that the rejection of the epistemological standpoint leads Dewey to adopt a much more contextualist explanation of the constitutive element than Lewis does. Other differences are much more substantial. Here I limit myself to analyzing one aspect which is distinctive of Dewey's approach to the matter.

As I have repeatedly remarked, Dewey holds that universal propositions provide partial definitions of the concepts used in inquiry. However, Dewey also rejects the idea that universal propositions are *true* by definition. Though it may seem paradoxical that something can be a definition and yet not being true, this is precisely Dewey's view. And far from being arbitrary and dispensable, that thesis can be easily seen to follow directly from some of his most ingrained assumptions: indeed, it is clear that since, by definition, propositions cannot be truth-bearers, universal propositions too can be neither analytically true nor analytically false.

In this sense, Dewey's approach is different from – and more original than – Lewis's one, which is committed to the classical view that definitions are analytically true. Here is what Lewis writes in this regard:

> The paradigm of the a priori in general is the definition. It has always been clear that the simplest and most obvious case of truth which can be known in advance of experience is the explicative proposition and those consequences of definition which can be derived by purely logical analysis. These are necessarily true, true under all possible circumstances, because definition is legislative [. . .]. Mind makes classifications and determines meanings; in so doing, it creates that truth without which there could be no other truth (Lewis 1929, 239–240).

In this passage, Lewis explicitly acknowledges that the content of experience does not play any role in justifying a definition: definitions are acts of choice, and insofar they are necessary and universal truths. In a Kantian fashion, he maintains that since the truth of a definition can be established in advance of experience, it is necessarily true, i.e., true under all possible circumstances. Definition, a priori, analiticity, and necessary truth are all essentially interconnected.

Dewey endorses the view that definition, a priori, and analiticity are intrinsically related, but he argues that these concepts do not imply that of (necessary) truth. In so doing, he emphasizes the postulational nature of the definitions formulated

[29]For a detailed analysis of the similarities between Lewis and Dewey's theories of a priori, please see Rosenthal (1987).

in universal propositions: as postulates, universal propositions cannot be said to be true or false in that they amount to free decisions by the inquirer about how to apply certain concepts. If, *within the context of a particular inquiry*, the inquirer decides that the character "having the parasite of the mosquito in the blood" is part of the definition of malaria, then it is not conceivable to get sick of malaria without having the parasite of the mosquito in the blood. But this is nothing but the statement of the stipulated use of the notion. Just as it does not make any sense to ask the truth-value of the rule that smoking is not allowed in public places, or to ask whether a hammer is true or false in its function of hammering a nail; similarly it does not make sense to ask whether universal propositions – whose function is that of reconstructing a problematic situation, and whose nature is that of stating how to use a concept – are true or false.

In so doing, by emphasizing the intermediate and instrumental character of universal propositions, Dewey's account – contrary to Lewis's one – is not challenged by Quine's attack to the analytic/synthetic distinction. At the same time, it offers a more fine-grained description of the various components that make up – to use Quine's vocabulary – the web of beliefs, thus providing the theoretical resources for a more refined understanding of the concrete dynamics of scientific practices.

4.5 The Construction of Objectivity

Within Dewey's framework, objectivity is *by definition* what is constructed in the judgment that ends inquiry. Being the result of a process of inquiry, that of object is a logical notion, in Dewey's technical sense. Recall Dewey's functional account of objectivity: "[t]he name *objects* will be reserved for subject-matter so far as it has been produced and ordered in settled form by means of inquiry" (LW12, 122). In the judgment the possibilities that are conveyed by the predicate are attributed to the subject as its potentialities. Stated in other words, once the judgment is formulated the inquirer stops reflecting over the propositions through which inquiry is carried on, and uses the content of those propositions in non-cognitive life-behaviors, thus modifying her habits of activity. As Dewey puts it in *Human Nature and Conduct*, objects represent "habits turned inside out" (MW14, 127).

Now, by treating objectivity as a logical notion, Dewey aims to achieve two distinct yet interrelated goals. On the one hand, he wants to rebut any metaphysical interpretation of the concept of objectivity, thus paving the way for a functional analysis of its fundamental features. On the other hand, he wants to articulate a radical form of scientific pluralism, which admits an indefinite number of ways of constructing scientific objects without for this reason having to give up realism.

Let's start with the first issue. As has been repeatedly remarked, one of the distinctive traits of Dewey's idea of the primacy of inquiry is that there is no need of a metaphysical backup in order for an element to carry out its logical function. Dewey's criticism here is explicitly directed against the traditional Aristotelian framework, to the effect that science recapitulates ontology. As Dewey reads it, what counts as substance according to the Aristotelian tradition is what is metaphysically robust and not subjected to change: properly speaking, only what is immutable is "fit to be subject [...] of 'true' propositions" (LW12, 133). On the contrary, in light of Dewey's functional perspective the notion of substance is to be understood in terms of the role it plays within inquiry: different substances have to be identified "in virtue of the different consequences which ensue from their operational use" (LW12, 132).

I am not interested in assessing the soundness of Dewey's interpretation of Aristotle's views; what is worth noting is, rather, the theoretical point that Dewey wants to make with that reading. According to the functional perspective that he endorses, being a substance means being something of a kind, and being something of a kind means being a knot in a net of inferential relations. The distinctive trait of substance is, therefore, its inferential stability and consistency, which grounds the capacity of the agent to deal with the objects of her environment in an adequate and successful way.

Dewey is explicit on this point. He writes that substance is "a logical, not an ontological, determination", and then adds, even more radically, that substantiality "is a form that accrues to original existence when the latter operates in a specified functional way as a consequence of operations of inquiry" (LW12, 131–32). Contrary to what is usually believed – namely, that for a classification to be satisfactory and effective it has to carve nature at its joints – Dewey argues that everything can be considered a substance provided that it enables the agent to draw inferences whose conclusions are predictively and explanatorily sound.

In this sense, the criticism that is traditionally directed against the idea of conventional kinds, and which lies at the ground of the distinction between natural and artificial kinds, can be shown to conflate two different lines of reasoning, one of which is correct while the other is – at least from Dewey's perspective – illegitimate. Clearly, Dewey is critical of any form of *conventionalism*, which he considers an inconsistent position, its inconsistency being due to its overtly intellectualist character. As repeatedly observed, Dewey is committed to the view that the *criteria* for adopting a certain classification do not amount to the decisions taken by the participants in the shared activity. The criteria are always logical – in Dewey's sense – and consist in the capacity of those classifications to reconstruct the problematic situation, when they are used in inquiry, or to be smoothly and continuously confirmed by the responses of the environment that they elicit, when they are used in non-cognitive activities.

Consequently, the metaphysical thesis that there are no *natural* kinds – or, alternatively put, that any natural kind is conventional – finds no room in Dewey's philosophy of science. This does not mean, however, that conventionalism is wrong all the way down. So, for instance, its insistence on the fact that classifications are always dependent on human interests and values is accepted by Dewey as a genuine component of his own account. Dewey's position can be read as an attempt to preserve that insight without accepting its metaphysical interpretation. In the last analysis, it revolves around the view that there is a difference between natural and conventional kinds, but such difference has to be cashed out wholly in inferential terms, instead of by focusing on their responsiveness – or lack of – to human values and interests.

Some classifications can be said to be conventional in the sense that they have been used at some time in the past, have proved themselves to be inferentially reliable – at least to a certain extent – and yet have turned out to be unsound in light of further research. What Dewey has in mind here are some past biological classifications: for instance, for quite a long time the traits of "having winds" and "capacity to fly" were taken as sound evidential signs of the kind "bird". So, under those assumptions there is a natural kind which encompasses insects, birds and bats. However, later scientific discoveries have shown that no natural kind of this sort exists; it was thus proved that the classification was wholly conventional, and that its plausibility was due exclusively to its inferential soundness: indeed, that principle of classification worked quite well only in the limited contexts of activity that were pursued at that time.

The goal of scientific investigation is, therefore, that of stripping off all the conventional elements from our classifications, thus paving the way for the discovery of genuinely natural kinds. As noticed, the distinctive feature of substantiality is the capacity of certain traits to "hang together as dependable signs that certain consequences will follow when certain interactions take place" (LW12, 132). Now, it seems reasonable to say that that capacity ultimately depends on how the world is. In this sense, there is nothing conventional in the fact that certain traits cohere with certain others, while others do not.

Dewey illustrates this point with the example of the common-sense substance "sugar":

> The quality sweet does not stand alone but is definitely connected with other observed qualities. As thus characterized, it enters into further situations in which it incorporates into itself additional qualifications. It is a sweet, white, granular, more or less gritty thing or substance, say, *sugar* (LW12, 131).

The fact that the quality 'white' hangs together with that of 'granular' does not depend on any convention. However, the fact of considering the trait 'sweet' as cohering with the other traits that constitute the substance 'sugar' clearly depends on the interests of the agents. Here is how Dewey formulates this insight:

> Sugar [...] is a substance because through a number of partial judgments completed in operations which have existential consequences, a variety of qualifications so cohere as to form an object that may be used and enjoyed as a unified whole (LW12, 131).

And, even more clearly:

> The conjoined properties that mark off and identify a chair, a piece of granite, a meteor, are not sets of qualities given existentially as such and such. They are certain qualities which constitute in their ordered conjunction with one another valid signs of what will ensue when certain operations are performed. An object, in other words, is a set of qualities treated as *potentialities* for specified existential consequences (LW12, 132).

Now, the operations to be performed are parts of more complex activities that it is up to the agents to decide whether or not to undertake. It is not difficult to imagine a scenario in which, either because of a different biological constitution or because of different social customs, a group has no interest at all in identifying sugar as a substance. In this scenario, no object of that kind would exist, since the members of the group would not respond in a differential and consistent way to the qualities that we associate to sugar.

The acknowledgment of the intrinsic relation of substantiality to the different contexts of activity and to the multiplicity of life-behaviors enables Dewey to provide a pluralistic and multifaceted account of objectivity which is in agreement with current scientific practices. Indeed, in many different fields of research scientists are concerned with types of object that do not have the ontological solidity traditionally attributed to natural kinds. Dewey's approach is malleable and flexible enough to adapt to the increasingly ontological variety of science.

Once again, the key notion is the emphasis on inferential stability as the only legitimate criterion for substantiality and objectivity. Now, inferential stability comes in degrees: an object can be more or less inferentially tied into a web of knowledge, and its degree of objectivity depends on its internal inferential articulation. Dewey is clear on this point: "[t]he greater the number of interactions, of operations, and of consequences, the more complex is the constitution of a given substantial object" (LW12, 132). In this sense, when an object is identified as one of a kind, its constitution is made complex by the fact that, thanks to the whole system of scientific classifications of which that kind is part, a great number of traits can be securely inferred from the observed ones. However, there are cases in which some traits are discovered to hang together in a sufficiently stable way – thus satisfying the conditions for objectivity – without being clear how they can be inserted in a system of inferential relations. In this case, the kind to which the object belongs is left greatly undetermined, and the inferences that can be drawn from its observable traits are quire narrow in scope and stability; nonetheless, it counts as a genuine substance, with specific potentialities that can be activated in some controlled courses of activity.

An example can be useful here to clarify this point.[30] Recent biological studies have been devoted to investigating the constitution of rocky intertidal biofilms. Biofilm is an extremely bizarre entity: etimologically, a biofilm is a thin coating comprised of living materia. In that field of research, biofilm (or Epilithic Microphytobenthos) is represented by the epilithon that forms on submerged rocks in streams and other water bodies, such as sea rocks (Wimpenny and Kinniment 1995, 99). Far from being a natural kind, biofilm of intertidal rocks is a mereological composition of prokaryotes and microbial eukaryotes like diatoms, protozoa and fungi, which produce many different extracellular polymeric substances (EPS) that keep all those elements embedded within a slimy extracellular matrix. This matrix provides the conditions for the survival of the microbial cells, and its properties depend on the

---

[30] I would like to thank Lorenzo Buccio for his expertise on intertidal biofilm and his extremely useful comments on this section.

extracellular polymeric substances produced by the cells composing the biofilm.[31] EPS immobilize biofilm cells and, by keeping them in close proximity, allow for their interactions.

In a case like that, one might be tempted to conclude that biofilm should not be considered a substance. In the last analysis, even though biofilms are ubiquitous in the environment, their composition is highly variable, to such an extent that both the proteins that constitute the extracellular matrix and the composition of the biofilm are often unknown. In addition, biofilm is constantly changing: it is an empirical hypothesis still under investigation that "under different conditions and/or at different times during the maturation of a biofilm, different components of the biofilm matrix may be of more importance to the integrity and function of the biofilm" (Karatan and Watnick 2009, 329). All these empirical discoveries seem to support a radical eliminativist approach to biofilm. According to that approach, biofilm does not really exist – it is not part of the furniture of the universe; at best, pragmatic legitimacy can be attributed to the *notion* of biofilm, since it actually plays an explanatory function in biological research.

However, an approach of this type seems too prescriptive with regard to concrete scientific practice. Indeed, biofilms are usually treated *in themselves* as complex biological systems, and it is widely acknowledged that, conceived of as a *mode* of life, biofilm is advantageous for microorganisms under a variety of conditions. In addition, rocky intertidal biofilm presents some traits that cohere together, thus making certain inferences possible. So, for instance, an important line of research is devoted to investigating what is called the biofilm architecture, thanks to which some interesting regularities among its determinants – medium composition (particularly the carbon source in the medium), presence of surfactants, various types of motility (flagellar, twitching, and swarming), and quorum-sensing effects – have been discovered.

Clearly, that biofilm does not count as a prototypical instance of objectivity and substantiality can be hardly questioned. Nonetheless, it is also difficult to deny that it satisfies many of the conditions that are usually referred to as criteria for objectivity. Furthermore, it is true that it performs an explanatory function, but it seems awkward to think it as a theoretical entity – among other things, it can be observed with naked eye and measured with various instruments.

To be clear. I do not want to take a position here regarding whether or not biofilm is to be accepted as part of the furniture of the universe. Neither am I willing to argue that a pragmatist philosophy of science is necessarily committed to the existence of entities of this kind. That decision is up to the researchers working in the field, and ultimately depends on the empirical discoveries that are and will be made on it, as well as on the use of the notion by the scientists. It does not – and cannot – depend on any philosophical argument in favor or against it. Mine is, therefore, a negative argument, whose aim is to get rid of those assumptions that aim to exclude that

[31] Different cells produce different EPS, and the latter, in their turn, affect and modify the properties and capacities of the matrix. On this point, see Flemming (2016) and Flemming et al. (2016).

option in an aprioristic way. There is a quite natural tendency to understand biofilm as a real entity, and that tendency is to be taken seriously.

When seen from this perspective, Dewey's functionalist and inferential approach to objectivity and substantiality appears particularly promising. By acknowledging the primacy of scientific practices, it can easily accommodate that demand of theoretical prudence. Moreover, its focus on inferential stability as the only legitimate criterion for attributing objectivity sets the stage for a pragmatic and contextual approach to ontological commitment. There might be prototypical instances of objectivity: however, these prototypical instances vary depending on the different contexts in which the agent is engaged. Prototypical instances of common-sense objectivity – an object (a table), a mereological sum (sugar), an event (a fire) – work perfectly well within common-sense activities, but they risk to be overtly restrictive when transferred in different and more complex life-behaviors, such as scientific inquiry.

The last remarks lead us directly to the issue of scientific pluralism.[32] Any scientific practice is recognized as free to construct its own objects as it fits to the needs of inquiry. As I read him, therefore, Dewey is a radical scientific pluralist who rejects both methodological and substantial unity of science in favor of an anti-reductionist image of science and scientific activity.

True, Dewey insists on the unity of scientific method, but it is clear that what has in mind is, rather, the unity of scientific *attitude*. It is this kind of unity that Dewey aims to represent with the idea of the unity of inquiry – the idea of a pattern of inquiry that is common to science and common sense – that he articulates in the *Logic*. The difference between attitude and method is formulated by Dewey in the following terms:

> The scientific attitude may almost be defined as that which is capable of enjoying the doubtful; scientific method is, in one aspect, a technique for making a productive use of doubt by converting it into operations of definite inquiry (LW4, 182).[33]

As is evident, Dewey does not argue for the unity of scientific method. As a technique for making a productive use of doubt, method is responsive of the specificity of the subject-matter under investigation. In this sense, there is no reason to assume that the methods successfully applied in a certain field of research will apply equally satisfactorily to other different domains. This is not just a prudential advice: it follows directly from Dewey's conception of inquiry, and, more precisely, from his idea of a normatively guided development and transformation of the

---

[32]The literature on scientific pluralism is wide. See, in particular, the so-called Stanford school of philosophy of science: Cartwright (1983, 1999), Dupré (1995), and the collection of essays (Galison and Stump 1996). A different approach to scientific pluralism, which centers on the notion of complexity, is (Mitchell 2003, 2009). For a detailed analysis of the notion of scientific pluralism, as well as of the various forms in which it may be articulated, see the classical (Kellert et al. 2006) and the more recent (Ruphy 2017). Dewey's scientific pluralism is different from those approaches in that it is primarily a pluralism about scientific activities.

[33]For a detailed analysis of the distinction, see (Campbell 1995, 100ff.).

original, problematic subject-matter. It also follows from his practice-centered view of science, and, in particular, from the thesis that any inquiry is individuated by the specific problem that it is called out to solve.

Something similar could be said about Dewey's criticism of the idea of reductionism, in which substantial monism is grounded. As John Dupré has remarked, "the most powerful source for the idea of science as a unified project is the commitment to substance monism: the belief that, in some important sense, the world is composed of some one kind of stuff" (Dupré 1995, 87). Now, it should come as no surprise that Dewey is highly critical of any approach of this kind. It would not be difficult to show that, on a strictly metaphysical level, substance monism flies in the face to Dewey's commitment to emergentism. However, since I have purposely and consistently avoided discussing Dewey's metaphysics throughout all this book, I won't dwell on it any further here. Similarly, I won't take into account the materialistic version of reductionism.

More interesting from my viewpoint is the rejection of reductionism on an ontological level. Since I will deal with Dewey's scientific realism and his account of scientific objectivity at length in the next chapter, I limit myself here to a preliminary remark, which sets the stage for further and more detailed analysis. Dewey defines object and objectivity as the form that subject-matter takes at the end of inquiry; consequently there is no need to assume – either as a working hypothesis or as a regulative idea – that some kind of reduction of the different scientific objects to a privileged set *must* be possible or *must* be pursued. A biologist deals with a certain object – say, biofilm – in a controlled and distinctive way, which means that she has habits of memory, prevision and action that fix the expected behavior of that object. The object is thus acknowledged to be one of a kind; certain potentialities are attributed to it, and certain consequences are expected to follow if some operations are carried out.

Within this framework, the reduction of an object or event to a set of allegedly simpler entities is not *necessarily* desirable: it might well be that the achieved reduction has negative effects on the capacity of a group of inquirers to solve the practical problems that they are trying to address. Once again, the decision in favor or against reductionism is to be taken at the level of the concrete inquiries that are carried out by scientists, and it might be piecemeal and contextual: reductionism may be useful in certain context and harmful in other.

Clearly, this is a restatement of the idea of the primacy of activity. In this case, however, the primacy which is thus stressed is of an 'ontological' kind: it amounts to saying that the ontological commitments of an agent or an inquirer depend on the specific demands of the activity – be it logical or extra-logical, scientific or practical – in which she is engaged.

Chapter 5
Realism

Abstract This chapter deals with the issue of scientific realism. Traditionally, Dewey's philosophy of science has been considered as a prototypical form of instrumentalism. On the contrary, I think Dewey is a scientific realist. My aim is to figure out what kind of scientific realism he embraces and defends in his texts. I reject the interpretation that Dewey's scientific realism is a form of structural realism. Relying on the conclusions reached in the previous chapters, I outline the main features of what I call Dewey's articulative realism, whose key assumption is the idea that the commitment to the existence of the entities postulated by our best scientific theories is a matter of the kind of activities that an agent is capable of undertaking.

Keywords Instrumentalism · Ontological commitment · Scientific objects · Structural realism · Structure · Connection · Relation · Language · Formulation · Articulative realism

5.1 Introduction

As is well known, pragmatism has been traditionally considered the prototypical form of scientific instrumentalism. The wrong idea that pragmatism recommends an anything goes approach to scientific explanation has led many interpreters to believe that it also endorses a deflationary approach to the existence of theoretical entities postulated by the theory. Put it roughly, if the ultimate criterion for the acceptance of any scientific theory is that it works, so the argument goes, then there is no need to be ontologically committed to the theoretical entities that are referred to in it.

Broadly speaking, instrumentalism holds that the unobservable entities postulated by science should be understood as instruments for the prediction of observable phenomena or for systematizing observation reports. More detailedly, an instrumentalist account of science may take two different forms. On the one hand, it can be grounded in the semantic thesis that there is no truth of the matter. Scientific theories are not true or false: since theoretical terms do not have meaning, the propositions in which theoretical terms are used are not truth-bearers; rather,

R. Gronda, *Dewey's Philosophy of Science*, Synthese Library 421,
https://doi.org/10.1007/978-3-030-37562-1_5

they are convenient ways of making predictions about observable entities. On the other hand, it can be grounded in the ontological thesis that the entities postulated by the best scientific theories do not actually exist. More precisely, instrumentalists advise us to resist the widespread and commonsensical assumption that one should be committed to the existence of the entities postulated by our best theories.

The goal of this chapter is to question the reading of Dewey's philosophy of science as a form of instrumentalism, by showing that he rejects both its versions.[1] It is Dewey himself who places himself in the realist camp as regards to the ontological status of scientific objects. In his reply to Reichenbach, who advances an instrumentalist reading of his philosophy of science, Dewey unquestionably rejects any allegations of being anti-realist about scientific entities. Nonetheless, it is indisputable that Dewey's realism is of a very particular type. In the following pages I will therefore try to outline the main features of his scientific realism, focusing on its semantic and practical components.

The chapter is structured as follows. The first section is devoted, firstly, to laying out Reichenbach's instrumentalist interpretation of Dewey's philosophy of science and, then, to reconstructing the arguments through which Dewey rejects that reading. In the second section I will present and discuss Godfrey-Smith's attempt to read Dewey's scientific realism as a form of structural realism. Finally, in the third section I will formulate my realist proposal. I will argue that Dewey's scientific realism is centered upon the idea of *expression* or *statement*, and that the attribution of reality to a scientific object is essentially interwoven with the capacity of the agent to use that object in a scientific practice.

---

[1] An instrumentalist reading of Dewey's philosophy of science has been advanced in Pihlström (2007) and in Shook (2002). The latter proposal is particularly interesting. In that text, indeed, Shook acknowledges that Dewey's pragmatic philosophy of science has room for propositions about postulated entities which are, by definition, not observable, but he also holds that "Dewey refuses to take a realistic attitude toward such 'objects'" (Shook 2002, 105). I will discuss and criticize Shook's proposal below. A neutral approach to the realism/instrumentalism issue has been defended by Hildebrand, who has argued that Dewey's attempt to avoid traditional philosophical dichotomies entails the dissolution of the problem of realism (Hildebrand 2003). A realist reading of Dewey's philosophy of science has been formulated by Rubeis in his recent essay *Beyond Realism and Antirealism? The Strange Case of Dewey's Instrumentalism*. I am strongly sympathetic with Rubeis's approach: I share his insistence on the importance of Dewey's practical standpoint, as well as his emphasis on the realist brand of Dewey's constructivism. I also agree with him that Dewey's realism is *unorthodox* since it is centered on the transformative function of activity (Rubeis 2017, 81). The two main points of disagreement between Rubeis's account and mine are that (a) I am more inclined than he is to distinguish between *logical* instrumentalism – the idea that logical elements are to be understood as instruments used to solve a problematic situation – and scientific instrumentalism – the ontological view, to the effect that we are not ontologically committed to the entities postulated by our best scientific theories; and (b) he is more inclined than I am to distinguish between ontological realism/antirealism and epistemological realism/antirealism (Rubeis 2017, 78).

5.2 The Reichenbach-Dewey Debate

In his contribution to the Schilpp volume on John Dewey, entitled *Dewey's Theory of Science*, Reichenbach offers a detailed reconstruction of Dewey's philosophy of science. Reichenbach's analysis is strongly sympathetic to Dewey's theoretical proposal: being committed to empiricism too, he acknowledges a strong similarity between his and Dewey's approach. Reichenbach appreciates Dewey's account of scientific method; he also praises Dewey's interest in the question of the justification of the principle of induction. Nonetheless, he considers the anti-realistic implications that he believes Dewey draws from his empiricism to be philosophically untenable. On this point the logical empiricist Reichenbach and the pragmatist Dewey parts their company.

Reichenbach's analysis of Dewey's anti-realist philosophy of science centers upon a passage from *The Quest of Certainty* in which Dewey seems to endorse a strong form of instrumentalism regarding scientific objects. The passage runs as follows:

> Water as an object of science, as H_2O with all the other scientific propositions which can be made about it, is not a rival for position in real being with the water we see and use. It is, because of experimental operations, an added instrumentality of multiplied controls and uses of the real things of everyday experience (LW4, 85).

Reichenbach reads Dewey as saying that H_2O does not refer to an existing entity in the world, but should rather be understood as an instrument that enables the agent to make reliable predictions about the behavior of the ordinary thing water. In this sense, he writes that Dewey attempts a "dissolution of scientific objects into relations of 'qualitative' objects" (Reichenbach 1939, 164).

Against the instrumentalist interpretation of the nature of scientific objects, Reichenbach deploys several arguments. Directed against different aspects of Dewey's proposal, those arguments aim to show that a probabilistic empiricism as the one developed by Reichenbach himself in *Experience and Prediction* – and which embodies several insights and ideas shared by Dewey himself (Reichenbach 1938, 80) – succeeds in developing a realist philosophy of science without relapsing into metaphysical realism. As Reichenbach puts it, the point at stake is to acknowledge that "probabilistic empiricism opens new ways for a realistic interpretation of scientific concepts not imperiled by any transgression into a domain of unknowable 'things in themselves'" (Reichenbach 1939, 177).

The first argument that Reichenbach puts forth is semantic, and is directed against the theory of meaning that Dewey shares with many other contemporary philosophers (Reichenbach 1939, 173). Both American pragmatism and European positivism, Reichenbach remarks, are nominalistic in that they stress the reducibility of scientific concepts to the things of the world as given to the senses. According to the nominalistic view, scientific terms work as abstract terms: just as an abstract term such as "the race of negroes" is reducible to statements about the individuals composing that class, the meaning of scientific concepts too can be formulated in terms of things of the "immediate world around us". In both cases, there is no need to assume that the concept refers to an independent and self-existent entity; rather, the abstractum is to be understood as "a kind of shorthand for groups of complicated statements about concreta" (Reichenbach 1939, 161). Peirce's pragmatic maxim, with its semantic insistence on the practical effects of scientific concepts; James's idea of the empirical cash-value of abstract notions; Dewey's statement that scientific object is an "instrumentality of multiplied controls and use of the real things of everyday experience"; and, finally, Mach's conception of physical things as complexes of elements given in immediate experience; all these philosophical proposals are considered by Reichenbach as different ways of articulating the fundamental instrumentalist insight, to the effect that scientific objects are completely reducible to statements about observable entities.

True, there is a difference between the two approaches, which Reichenbach is ready to emphasize. While positivists aim to provide a reduction of both scientific and commonsense objects to sense data, American Pragmatism in general – and Dewey's philosophy in particular – has the merit of "insist[ing] upon the idea that sense data are abstractions as much as are objects of physical science": sense data are not the material out of which primary experience is composed, but are refined products of reflection, whose field of applicability is determined by the scope of the problems that they are expected to address (Reichenbach 1939, 162). Reichenbach acknowledges that the idea of the primacy of experience over sense data represents a relevant philosophical acquisition. However, no matter how philosophically relevant this aspect may be, he believes that Dewey's instrumentalist philosophy of science is untenable.

In order to clarify this point, Reichenbach deploys the theoretical apparatus elaborated by logical empiricists, and in particular their linguistic approach to philosophical questions. In the wake of Carnap, Reichenbach highlights that "questions concerning the existence of things assume a form more easily to be answered if they are turned into questions concerning the meaning of propositions" (Reichenbach 1939, 174). In light of the linguistic framework of analysis, Dewey's instrumentalism can be reformulated as the thesis that "every scientific statement is equivalent to statements about observable concrete objects" (Reichenbach 1939, 174). What is important to remark here is that, according to Reichenbach's reading, Dewey's theory of meaning counts as a sample of the *verifiability theory of meaning*: the distinctive feature of this view – which is the kernel of the operationalist analysis of physical concepts – is that the inference from observational statements to scientific statement is analytic; the two sets of statements are logically equivalent. As Reichenbach puts it, "the logical background of Dewey's theory of instrumentalism of scientific concepts" can be formulated as follows: "[a] proposition has meaning only in so far as it is verifiable, and an abstract statement has therefore only so much meaning as has the group of concrete statements to which it refers" (Reichenbach 1939, 174).

Against the verifiability theory of meaning, Reichenbach remarks that the inference from a set of observational statements to a scientific statement in which the corresponding theoretical term is employed is not analytic, as operationalists believe, but synthetic. Reichenbach frames this point in terms of the concept of probability: a scientific statement adds something new to the observational premises from which it is derived. In particular, the scientific statement is applicable to an unlimited class of predictions of new facts, which is neither identical nor reducible to the original set of observational statements. In this sense, the idea of a logical equivalence between sets of observational statements and scientific statements is wrong.

The second argument against Dewey's instrumentalism has to do with the role of inferential processes in the construction of scientific objects. Reichenbach notices that there is a strong continuity between the inferential processes that are used to discover or postulate scientific objects and those that are used in our everyday transactions with ordinary things. Take, for instance, a bent stick in water: in this case, the qualitative object that is actually perceived is not considered to be real, but is replaced by an inferred object – a straight stick – which is seeable under different conditions. The inferred object is, thus, assumed to be the real object; which means that there are uncontroversial cases in which "primitive experience compels us to abandon the perceive objects and to replace them by inferred objects the reality of which is better founded than that of the perceived objects" (Reichenbach 1939, 165). So, if these inferential methods are relied upon in the latter case, there is no reason to question their validity when they are applied in scientific investigations, with the aim of postulating inferred, real objects that explain the behavior of the perceived, qualitative objects.

The third argument is related to Dewey's alleged resistance to admit scientific entities, which Reichenbach attributes to his fear that their introduction could lead to some sort of depreciation of the reality of ordinary things. Reichenbach counters such concern by noticing that the attribution of reality to scientific objects does not necessarily end up with the attribution to ordinary things of the malicious status of appearance. The problem is purely terminological: appearance can be used in two different senses, which have to be kept separate. There is nothing wrong in the pre-philosophical and completely harmless use of appearance: "[t]he fisherman who dips his oar into the water and sees it bent", Reichenbach writes, "does not turn transcendentalist if he says that this is only appearance, and that in reality his oar is not bent" (Reichenbach 1939, 168). It is enough to stick to this sense of the term 'appearance' to get rid of any kind of philosophical trouble. Now, the same approach should be adopted when dealing with the relation between scientific and ordinary objects. Even in this case, indeed, the fact that, upon reflection, an ordinary object turns out to be something radically different from how it appears to perception does not entail that its perceived features are to be considered appearance. "The discrepancy of the seen bent oar and the inferred straight one is repeated, on a higher level", so the argument concludes, "in the discrepancy between the seen continuous nature of its wood and the inferred atomistic pattern: in both cases the objective thing is different from the seen subjective thing" (Reichenbach 1939, 169).

The fourth argument is directed against Dewey's thesis of the unamendability of perception as a ground for inferring the unreality of scientific objects. Reichenbach correctly attributes to Dewey the view that perception is a natural transaction of an organism with its environment. When seen from this perspective, Dewey argues, it makes no sense to say that the bent stick in water is less real than the straight one since both are different ways in which an object may appear; they both depend on the laws of refraction of light rays. Accordingly, both the perceptions are biological phenomena, and, for this reason, it is inappropriate to treat them differently as regards to their reality. Similarly, it is inappropriate to analyze them in epistemological terms: it is wrong to consider the bent stick a wrong or mistaken perception, and the straight stick a true one. In support of this consideration, Dewey stresses that a photographic camera would behave like the human eye, and would furnish the same report under the same conditions. This implies that there is no error for which perception should be held responsible: the language of truth and falsity simply does not apply in the case under consideration.

Reichenbach rejects this approach. There is a sense, he argues, in which a photographic camera too could be said to make mistakes. Reichenbach distinguishes between two different ways in which a picture taken by a photographic camera can be used: it can be interpreted as an observational report of the stick or it can be interpreted as the report of the projection of the stick on the film. In the second cases, the picture is correct and reliable, whereas in the first case it can be held responsible for a mistake in that it provides unreliable information about the object that it depicts. Consequently, the same insight can be extended to the analysis of perception. Our senses may be considered as registering instruments that function reliably under certain conditions. On the contrary, when senses are employed under non-normal conditions, their reports are unreliable. Properly speaking, however, to be unreliable does not mean to be accountable for a mistake; some further premise is needed to justify that conclusion. Now, according to Reichenbach, senses can be said to make a mistake on the basis that they lack the ability, to use his own words, "to correct the seen objects in correspondence to the abnormal physical conditions in which the object is presented" (Reichenbach 1939, 171).

Those remarks are used by Reichenbach as a springboard to challenge the pivotal Deweyan idea of the reality of tertiary (and secondary) qualities. As has been highlighted in Chap. 3, Dewey is committed to the thesis that tertiary qualities are objective properties of a situation as a whole: indeed, this thesis is central to Dewey's philosophy of science in that tertiary qualities provide the normative guidance that is necessary for an inquiry to be successful in reaching its conclusion. Reichenbach charges Dewey with conceptual confusion. We should distinguish, Reichenbach suggests, between two senses in which the conceptual couple subjective/objective can be employed. It is clear that a natural phenomenon or event such as perception can be investigated in a scientific and objective fashion. In this sense, he writes, "though psychology deals with subjective things it is objective because it states that persons of a certain type under certain conditions observe these subjective things" (Reichenbach 1939, 179). Nonetheless, the fact that secondary and tertiary qualities can be successfully accounted for in objective terms does not imply that they should be considered objective in themselves. It is the latter thesis that Dewey upholds. But, Reichenbach concludes, there are strong reasons in favor of the opposite view.

Reichenbach reasons as follows. Suppose for a moment that secondary and tertiary qualities are qualities of things – I have argued in Chap. 3 that this is not Dewey's position on that issue, but this is not relevant for the case. If this were true, it would follow that they should be objective properties of external things. However, it is platitudinous that secondary and tertiary qualities are not considered as objective properties of things in the same sense in which primary qualities are. Secondary and tertiary qualities are qualities of things only for those persons or organisms whose bodies possess certain specific organs: "[t]he things of the colour-blind man have no colours, but show only shades of grey [sic] in all variations" (Reichenbach 1939, 178). Consequently, secondary and tertiary qualities are *subjective* qualities.

There is, so the argument runs, a conceptual confusion at the basis of Dewey's belief that tertiary qualities are objective. According to Reichenbach, Dewey does not realize that it is incorrect to infer from the true premise

A) within the scientific image of the man in the world, tertiary qualities are explained as objective *relations* between an organism and some parts of its environment which the former is sensitive to;

to the false conclusion

B) within the manifest image, tertiary qualities are objective *qualities* of things.

But this is clearly a mistake. "Within [the] interpretation as objective", Reichenbach points out, "the secondary and tertiary qualities are no longer *qualities* of things, but *relations* between things and observer" (Reichenbach 1939, 180). Now, since within the manifest image – what Reichenbach calls here "our immediate world" – secondary and tertiary qualities are treated as *qualities* and not as relations, they cannot be considered objective. Properly speaking, objective qualities are only those that can be explained without having to introduce the observer into the description. In reality, objective is what is real, and what is ultimately real are the scientific objects that are inferred from immediate existence.

In different ways and through different means, the four arguments laid out by Reichenbach are meant to converge in showing that the anti-realist concerns that Dewey displays in his work are unfounded: there is no need to hold – this is Reichenbach's point – that the ascription of reality to scientific objects entails the depreciation of ordinary objects or that it leads to some sort of metaphysical or epistemological dualism. There is no need, therefore, to hold that scientific objects are to be dissolved into relations of qualitative objects. It is possible to preserve – through the notion of immediate existence – the common-sense insight that our world is populated with subjective *qualities*, and, at the same time, attribute reality to scientific objects. This is the goal that Reichenbach's probabilistic empiricism purports to achieve.

As is evident, what all these arguments have in common – apart from a serious misunderstanding of Dewey's views – is the assumption that Dewey's interpretation of scientific objects is strongly anti-realist, and that its distinctive thesis is the reduction of scientific objects into relations of qualitative objects. It is that assumption that Dewey rejects in his dense reply to Reichenbach, contained in *Experience, Knowledge and Value: A Rejoinder*.

In that text, Dewey outspokenly rejects Reichenbach's instrumentalist reading of his philosophy of science. Referring to the contemporary debates in the field, he writes: "Duhem [...] many years ago presented a view which amounted in effect to saying that scientific objects are symbolic devices for connecting together the things of ordinary experience. Others have held that they were devices for facilitating and directing predictions" (LW14, 22). Now, compared to these views, Dewey explicitly claims that his position is less radical: "my view does not go as far as these", he comments (LW14, 23).

But what is the theoretical kernel of Dewey's position? And how does his philosophy of science look like? Clearly – and Dewey agrees with Reichenbach on that – the problem that instrumentalism purports to address and solve with the thesis of the unreality of postulated scientific objects is genuine: understanding the relation between scientific and ordinary objects is *the* problem that the rise of modern science has created and put on the agenda of philosophy. Dewey is also willing to take a further step: indeed, he is ready to admit that the rationale behind instrumentalism is substantially correct. The conflict between scientific and ordinary objects should be defused, and the only way in which this goal can be achieved is by showing that the two kinds of entities are by no means rivals. What goes wrong with instrumentalism is that the strategy through which it tries to reach that goal implies withholding the commitment to the reality of scientific objects. Dewey wants the best of both worlds: his aim is to preserve both the commitment to the reality of scientific objects and the relaxing idea that there is no conflict and rivalry between the latter and the ordinary objects of our primary experience.

Dewey's proposal revolves around two assumptions. The first one has to do with his realistic theory of relations; the second concerns the proper function that has to be attributed to scientific objects. I deal with the latter assumption first; the former needs a more careful and thorough analysis that will occupy the rest of the chapter.

The starting point of Dewey's analysis is the rejection of Reichenbach's reconstruction of the relationship between ordinary and scientific objects. As has been highlighted above, Reichenbach insists on the distinction between appearance and reality, and he does so for the purpose of making room for a realistic interpretation of scientific objects. He maintains that scientific activity is to be understood as the process through which inferred entities are substituted to ordinary objects: there is a continuity between scientific inferences and those common-sense processes of reflection through which, for instance, the perceived bent stick is substituted with a non-perceived, inferred stick. It is that continuity which guarantees the reality of postulated scientific entities.

Dewey criticizes the way in which Reichenbach formulates the relation between the two kinds of objects, from a twofold point of view. First of all, he notices that the goal of the postulation of scientific entities is not that of substituting ordinary objects with other, more real entities, but rather that of *explaining* their behavior: "[w]hat science does", Dewey remarks, "is not to correct the thing of ordinary experience by substituting another thing but to *explain* the former" (LW14, 24). And explanation is always and necessarily explanation *of* something real. True, Dewey concedes that his use of real in this context might look idiosyncratic, and that Reichenbach might be right in saying that he had better employ "existential" rather that "real" to refer to the fact that phenomena such as dreams, insanity, and fantasy are natural products originating from the transactions of an organism with the environment. Nonetheless, all this conceded, it remains indisputable for Dewey that the *explanandum* must have substantial "existentiality" and stability: if it were not so, explanation would not be possible.

The search for an explanation is the search for some sort of correlation between relevant phenomena. In the case of the perception of the bent stick, a satisfactory explanation is that which states the correlation between "changes in the density of media and changes in the refractive index of light". And it is worth stressing, Dewey remarks, that it is the ray of light itself which is bent: contrary to what Reichenbach says, "there is no illusion or 'appearance' about *that*" (LW14, 24).

The latter remarks highlight another relevant point of disagreement between Dewey and Reichenbach. From Dewey's perspective, Reichenbach tends to over-intellectualize immediate experience; the very idea of substitution is grounded in the assumption that ordinary and scientific objects are substantially identical in their function and scope of application. Dewey challenges this approach by noticing that there is no need to resort to scientific knowledge in order to account for the "substitution" of the bent stick with a straight one. The right way to put the whole issue is in practical terms, tracing it back to the concrete activities that an agent is capable of undertaking with the object. Dewey formulates this insight in the following terms:

> It is not *scientific* knowledge which substitutes bending of light, as it passes from one medium to another, for bending of the stick. For "bent" and "crooked" in objects of perception are matters of motor adaptive responses. I doubt if any oarsman ever failed to make the correct response, the response being "correct" because it produces the consequences intended in the act of using the oar. The boatman, then, never supposed the *oar* was bent. A fisherman who catches fish by spearing them requires but a few trials, with no help from "science," to form a habit that is expressed in effective motor adjustments (LW14, 24).

I think this passage is to be read as implying two different things, which are closely related to two distinctive features of Dewey's philosophy of science. First of all, it is wrong to take the deliverances of senses out of their proper context of production, and to treat them as elements whose only function is that of faithfully depicting the things that have caused them. The bent stick in the water is a stimulus that asks for a proper response; and part of the understanding of the significance of a stick consists in the capacity to deal with those contexts of activity in which it appears bent. In this sense, what is wrong in Reichenbach's approach is the epistemological perspective from which the phenomenon of perception is investigated.

Secondly, I read that passage as meaning that philosophers should be careful not to flatten the structural and functional differences between primary and secondary experience. Recall what has been said in Chap. 1. In that chapter it has been shown that scientific life-behaviors are characterized by the search of knowledge for the sake of knowledge; common-sense life-behaviors, on the contrary, are concerned with the use and enjoyment of ordinary things, as well as with solving practical problems through 'practical' reflection. Dewey's point is that the two kinds of life-behaviors have different purposes and, consequently, different criteria of satisfaction. It is not only that the perception of the bent stick functions perfectly well in the context of primary experience.[2] It is also that the corrective function of the adaptive motor responses in cases of ordinary experience – those motor responses that enable the oarsman to appropriately use the oar in water – is essentially different from the corrective function performed by "controlled and systematic *inference* in cognitive experience of the scientific type" (LW14, 27). This is an aspect of the relationship between science and immediate experience that Reichenbach completely overlooks.

In this sense, there is no possible conflict between scientific and ordinary objects. Neither is the reality of scientific objects somehow denied or called into question by acknowledging the different functions performed by ordinary and scientific objects. Indeed, Dewey holds that the function of scientific explanation is to expand the web of relations in which an element can be involved. As will be shown in a moment, scientific entities are understood by Dewey as statements of the *connections* between different sets of changes, within a controlled, experimental context: they enhance the capacity of prediction and control of the behavior of ordinary objects. Dewey is explicit on this point. It is worth quoting the passage in its entirety:

> The foundation of [Reichenbach's] criticism is the belief that my identification of the scientific object with *relations*, instead of with some kind of existing non-relational things, commits me to the doctrine of the "*non-reality*" of scientific objects. This point is so fundamental that I am grateful to Dr. Reichenbach for the opportunity to discuss the matter. For I certainly have never intended to say anything which could lead directly or indirectly to a belief that I hold a "non realistic interpretation of scientific concepts." On the contrary [...] the actual operative presence of *connections* (which when formulated are *relational*) in the subject-matter of direct experience is an intrinsic part of my idea of experience (LW14, 20).

Among the other things, this passage shows beyond any reasonable doubts that Dewey *believes* that his own position is not open to the criticism of being instrumentalist. However, Dewey's statement is hardly self-evident – to such an extent that Reichenbach reaches the opposite conclusion from the same premises. It has, therefore, to be somehow justified.

---

[2]See, for instance, the following passage: "What lies back of it [the denial of the validity of the view that scientific and common sense objects are rival] is the belief that the qualitative traits of the things of ordinary common sense knowledge are not only legitimate but necessary in connection with one kind of problems, – those of use and enjoyment –, while the so-called 'conceptual' objects of science are legitimate and necessary for the kind of problems with which scientific inquiry is concerned. Hence they are not rival claimants for occupancy of the seat of 'real' knowing; and one does not duplicate in 'true' or objective fashion what the other presents in a merely apparent and subjective fashion" (LW14, 21).

Let's look at this point in more detail. Reichenbach draws the conclusion that Dewey's theory of science is a form of instrumentalist from the premise that Dewey *identifies* the scientific object with a set of relations. Now, there is no doubt that Dewey endorses the latter view: he explicitly states that scientific object is nothing but the correlation between different sets of changes – in the case of the bent stick, between the "changes in the density of media" and the "changes in the refractive index of light". So, the premise is correct. Nonetheless, Dewey rejects the conclusion: "I certainly have never intended to say anything which could lead directly or indirectly to a belief that I hold a 'non realistic interpretation of scientific concepts'" (LW14, 20).

The reason why Reichenbach is prompted to jump to that conclusion is because, according to Dewey, he is committed to a too narrow conception of what a sound naturalistic empiricism should be. It is true that pragmatists and logical empiricists share the same critical attitude towards the idea that things should be reduced to sense data. Dewey and Reichenbach agree on this. However, Dewey believes Reichenbach to be still attached to what he calls "particularist empiricism" – namely, the view that "'relations' have not the empirical reality possessed by things and qualities" (LW14, 21).[3]

Clearly, Reichenbach is far from being alone in this thought. Rather the contrary, the idea that 'relations' are less real than the elements that they connect is widely diffused and accepted not only in the history of philosophy but also in the history of science. Take, for instance, psychology. The downgrading of relations lies at the basis of the mainstream approaches to psychological inquiry during eighteenth and nineteenth century: both traditional empiricist associationism and idealistic spiritualism shared that framework of analysis. It was William James the first who argued for the theoretical availability of a different approach, with acknowledges the same degree of reality to relations and things (LW10, 125).

The point that Dewey wants to make is that James's insight can be generalized. Recall here what Dewey says about experience as a double-barrelled word, which includes both what is done and how it is done. It is part of Dewey's idea of experience that "*actions* or modes of actions, ways of operating, are also contained in what is directly experienced" (LW14, 20). Stated in other terms, the ways in which an object responds to the actions performed on it by an agent have the same ontological status as the qualities that are considered constitutive of that very object. In light of Dewey's practice-oriented approach, there is no solid ground to draw a clear-cut *ontological* distinction between qualities and 'relations'.

I will come back to this issue later. For now, I would like to stress a more general aspect of Dewey's argument. It is clear that if 'relations' are real – as real as things or qualities are – the identification of scientific objects with relations does not directly

[3]Please notice that Dewey puts the word 'relation' in quotes so as to stress that he is adopting Reichenbach terminology without accepting it. I will follow Dewey's use while discussing the Dewey-Reichenbach debate. The reason why Dewey puts the term 'relation' in quotes is due to the fact his and Reichenbach's use of the term 'relation' differ consistently. I will come back to this in Sect. 5.4. For now, please take the notion of relation in its traditional philosophical sense.

and immediately lead to the conclusion of the unreality of scientific objects; there is no need to hold that they should be boiled down to fictions whose validity is determined exclusively in terms of their predictive capacity. Once it is admitted that the thesis of the reality of scientific objects does not depend on the controversial assumption that only things and qualities are real, room is made for a different way of framing the issue of scientific realism. But what does it mean that scientific objects *are* relations? What kind of ontology does this identification support?

In the following section I will explore a possible way of making sense of those theses by assessing whether or not Dewey's philosophy of science can be consistently understood as a form of structural realism. Then, in the final section of the chapter I lay out my account of Dewey's scientific realism, which is centered on the functions performed by scientific objects within scientific life-behaviors.

5.3 Is Dewey a Structural Realist?

The idea that scientific objects are relations challenges our traditional way of conceiving the kindred notions of objectivity and reality. The standard view – at least in the case of *external* relations – is that relations hold between elements whose reality is independent of their being involved in those relations. Aristotle is Alexander's teacher: the relation "being the teacher of" holds between two individuals – Aristotle and Alexander – who are characterized and distinguished by a series of monadic properties which are more fundamental than the relational ones. It is because Aristotle is a man, he is learned, etc., that he is Alexander's teacher: the relation of "being the teacher of" supervenes on the properties of its relata.

As Godfrey-Smith has pointed out in *Dewey and the Subject-Matter of Science*, Dewey is critical of the standard view and the downgrading of the reality of relations that it implies. Godfrey-Smith is undoubtedly right on this: Dewey's theory of inquiry, as well as his conception of cognitive meaning, is grounded in the idea that *only* relations can be properly known. Indeed, the qualitative features of the things composing the world are beyond knowledge. Not in the sense of being unknowable, but rather in the sense that they are not a matter of knowledge: in Dewey's terminology, they are had, used, enjoyed, suffered. In the "non-cognitive side of experience", as Godfrey-Smith calls it, we encounter and have to do with *qualities*, while in what may be called the cognitive side of experience – following Godfrey-Smith's trail – we are concerned exclusively with *relations*.

Here is how Godfrey-Smith formulates the issue:

> For Dewey, relations are known and qualities are "had". Neither is primary in a metaphysical sense, more real or more furniture-like. Neither is to be "explained away" in some manner that is not applicable to the other. They have different and complementary roles in our lives. The error of downgrading relations is rectified without making the error of installing relations in the position of metaphysical primacy once occupied by qualities (Godfrey-Smith 2011, 77).

I like this way of putting the matter: it is a clear but concise presentation of Dewey's views on the topic. Now, the reason why such presentation is relevant to the issue under discussion is that it highlights the similarities existing between Dewey's philosophy of science and the structuralist tradition that originates in the works of Poincaré, Russell and Cassirer, and finds its mature formulation in Worrall's *Structural Realism: The Best of Both Worlds* (Worrall 1989).[4] That strong similarities exist between Dewey's position and those of many of his contemporaries is far from surprising: there is no doubt, indeed, that Dewey shares many of the fundamental assumption of the structuralist tradition. Accordingly, Godfrey-Smith's proposal to assess whether Dewey's philosophy of science, with its insistence on the theoretical centrality of the notion of relation, can be understood as a form of structural realism is worthy of examination.

I will proceed as follows. Firstly, I will very briefly sketch the fundamental tenets of contemporary structural realism. Then I will turn attention to Godfrey-Smith's account of Dewey's position, and I will analyze his argument against the assimilation of Dewey's philosophy of science to structural realism, either in its epistemic or ontic form. Finally, I will criticize a few points of his reconstruction, with the aim of highlighting those idiosyncratic aspects of Dewey's scientific realism that make the latter's use of the notion of structure different from the standard versions of structural realism. The acknowledgment of such difference sets the stage for the analysis of Dewey's philosophy of science as a form of articulative realism, whose main features are developed in the final section of the chapter.

Contemporary philosophy of science is characterized by a tension between two opposite and yet equally legitimate insights, namely, the no miracles argument – which supports a realist view of science – and the pessimistic meta-induction – which, on the contrary, suggests an antirealist approach.[5] Structural realism is the most successful attempt to combine those two perspectives on the nature of science into a single comprehensive view. The idea behind this position is that it is possible to preserve a strong form of realism – thus satisfying the demands of the no-miracle argument – without committing yourself to the apparently untenable thesis that science has already delivered a true description of the world. An equilibrium between the two insights can be achieved by assuming that there is a strong and significant degree of retention between a theory and the one that it replaces, but that such retention has to do with the structures that are expressed by the mathematical apparatus of the theories.

Structural realists say something of this kind: it is very likely that the theories that we now hold as true commit ourselves to the existence of scientific entities that do

[4]In recent times, much has been written on the history of structural realism. See, among the others, Zahar (2001), Chapter 2, and Brading and Crull (2017) (on Poincaré); Massimi (2010b) and French (2014), Chapter 4, (on Cassirer); and Demopoulos and Friedman (1985) (on Russell).

[5]As is well known, the no-miracle argument has been originally formulated by Putnam in (1975, 73). The pessimistic meta-induction, on the contrary, goes back to Laudan (1981). The bibliography of the debate that has arisen about the validity of these two arguments is so extensive that it cannot be summarized here.

not exist. Nonetheless, the mathematical expression of the relations existing in the physical world seems to be much less open to the possibility of being wrong. This idea traces back to Poincaré's reconstruction of an important episode in the history of science, that is, the so-called Fresnel-Maxwell case. The point that Poincaré wants to stress is that, though Fresnel and Maxwell hypotheses refer to different entities – so that it can be argued that their ontological commitments are in conflict the ones with the others – the basic equations of the two theories are approximately the same. So, from Maxwell's perspective, Fresnel completely misapprehended the nature of light and yet his theory accurately described its structure and its observable effects.

Poincaré is quick to extend that historical remark into a general theory about the nature of science. In his book *Science and Hypothesis* (1905) he explicitly states that "the aim of science is not things themselves [...], but the relations between things"; and then he is willing to take a step further and add that "outside those relations there is no reality knowable" (Poincaré 1905, xxiv). If rightly understood, science turns out to be a reliable source of information on the *structural* features of reality, not on its intrinsic ontological properties.

Something very similar, though much more philosophically refined, was claimed by Russell in *The Problems of Philosophy* (1912) and in *The Analysis of Matter* (1927). Russell's version of structural realism is grounded in two fundamental psychological theses, namely the idea that different effects imply different causes (the so-called Helmholtz-Weyl Principle) and the assumption that relations between percepts have the same logico-mathematical properties as the relations between their causes. Let's take these premises as true: it follows that it is impossible for us to have knowledge of the non-perceptual causes of our percepts; however, because of their structural similarities we can nonetheless achieve knowledge about the structural properties of the objects that cause our percepts.[6]

The general conclusions that Russell draws from those remarks are close to Poincaré's views. Russell too argues that science is concerned exclusively with the knowledge of the relations that hold between non-perceptual objects; he also maintains that we can have knowledge of all the properties of the structural relations that are revealed to us by our percepts. On the contrary, the physical objects that cause the latter "remain unknown in their intrinsic nature" (Russell 2001, 17). This entails that the intrinsic qualities of physical objects are constitutively beyond the scope of our knowledge. Accordingly, structural realism is a plea of humility: the intrinsic nature of the physical world – whose relations are expressed by the mathematical structures of a theory – remains hidden.

In recent times, structural realism has become more internally variegated. A distinction between ontic structural realism (OSR) and epistemic structural realism (ESR) has been advanced, which has dramatically altered the shape of the debate. In light of this conceptual distinction, Russell and Poincaré's approaches count as

[6]It is worth noting that serious doubts have been expressed about the capacity of the Helmholtz-Weyl principle to support structural realism. See, in particular, Psillos (2001, S15).

somehow prototypical forms of ESR.[7] Indeed, ESR is characterized by the epistemic thesis that we are justified in adopting a realist attitude towards structures, while, at the same time, we leave open the possibility that there is something more in the world than mere structures.[8] OSR is a stronger and much more controversial position: it states that what there is in the world are structures all the way down. The reason why we cannot know the intrinsic nature of the physical world but only its structures is, therefore, that there is nothing like that.[9]

Obviously, this broad-brush sketch of contemporary structuralism is far from being accurate or barely satisfactory. It is nonetheless good enough for my purposes. As said, Godfrey-Smith has suggested taking into consideration the possibility that Dewey's realism is to be interpreted as a form of structural realism. Now, if one pays attention to the following passages from *The Quest of Certainty*, the similarities between structural realism and Dewey's philosophy of science to which Godfrey-Smith refers seem undeniable:

> 1) It is unnecessary that knowledge should be concerned with existence as it is directly experienced in its concrete qualities [. . .]. What science is concerned is with the happening of these experienced things [. . .]. Its aim is to discover the conditions and consequences of their happening. And the discovery can take place only by modifying the given qualities in such ways that relations become manifest [. . .] these relations constitute the proper objects of science as such (LW4, 84).
>
> 2) there is one common character of all such scientific operations which it, is necessary to note. *They are such as disclose relationships* [. . .]. Qualities present themselves as just what they are, statically demarcated from one another. Moreover, they rarely change, when left to themselves, in such ways as to indicate the interactions or relations upon which their occurrence depends [. . .] the aim of inquiry is to correlate events with one another. Scientific conceptions of space, time and motion constitute the generalized system of these correlations of events (LW4, 100–101).

All these similarities notwithstanding, however, Godfrey-Smith denies that Dewey's philosophy of science should be considered as a form of structural realism – at least, as the latter is usually understood. If Godfrey-Smith is right, the possibility is left open that emphasizing the theoretical and explanatory primacy of the notion of structure or relation is not enough for a scientific realism to qualify as a form of structuralism realism. This is precisely the point that I would like to make.

Let's start with ESR. Godfrey-Smith believes that this version is the one that comes closer to Dewey's views. I do not agree with him on this point, since a reading along this line is at risk of giving an epistemological twist to Dewey's philosophy of science that, in my view, is not only incorrect, but also deceiving. In any case,

---

[7] See, for instance, Votsis (2004).

[8] The literature on the topic is considerable. See, among the others, Worrall (1989) and Morganti (2004, 2011).

[9] OSR has been originally introduced by French and Ladyman, and then developed into a philosophical proposal that aims to provide a global framework for the understanding of contemporary physics, and, in particular, quantum mechanics. See, among the others, Ladyman (1998), French and Ladyman (2003, 2010) and French (2014).

leaving this issue aside for the moment, Godfrey-Smith is right in stressing that some relevant differences exist between the two approaches, which impede any identification of the two.

In particular, he remarks that the reason why ESR looks appealing to a realist is that it provides an antidote against the anxiety about our possibility to have access to the things that lie behind, and sustain, the relations that we formulate as structures in our theories (Godfrey-Smith 2011). Now, Dewey's recourse to the notion of structure does not seem to be motivated by such concern. As Godfrey-Smith rightly points out, for Dewey the distinction between intrinsic qualities or properties of things and their relations is not ontological, but functional (Godfrey-Smith 2011, 78). And in support of this reading he quotes a passage from *Experience and Nature*, in which Dewey maintains that it is true that "[t]hings in their immediacy are unknown and unknowable", but this does mean that they are"remote or behind some impenetrable veil of sensation of ideas", but simply that "knowledge has no concern with them" (LW1, 74).

I think Godfrey-Smith is right on this point. This said, I believe that something more can be said about the differences between Dewey's realism and ESR. I add two further remarks on this issue. Firstly, Dewey does not accept the Helmholtz-Weyl principle – or the so-called sign theory (*Zeichenlehre*) of perception – that lies at the core of ESR. Traditionally, this view is associated to Helmholtz, and is a relevant aspect of the nineteenth century "Go Back to Kant" movement. According to Helmholtz, perceptions should be understood as signs which refer to their causes without representing them. And it is only because there are regular connections between perceptions and their causes that we can use our perceptions to infer the properties of the things that lie beyond the realm of appearance (see MW7, 139–140).

Dewey rejects this account of perception, for several reasons. First of all, Dewey endorses a naturalistic and ecological theory of perception; now, such approach is clearly at odds with Helmholtz's account. Secondly, the causality that lies at the ground of the sign theory of perception – at least as the latter is formulated within an ESR framework – is a relation that is not empirically accessible, rather than a naturalistic event. Thirdly, and most importantly, the idea that the causal relation between objects and our perceptions of them is not empirically accessible opens the door to a metaphysical analysis that Dewey does not want to undertake. The latter point leads to the second remark that I would like to make about the differences between ESR and Dewey's philosophy of science.

One of the distinctive features of Dewey's philosophy is the rejection of the very idea of things-in-themselves. And it is not by chance, therefore, that he identifies that issue as one of the substantial points of conflict between pragmatists and their critics. So, for instance, in *A Short Catechism Concerning Truth* (1910) Dewey writes:

> the opponents of pragmatism have been forced by the exigencies of their hostility to resuscitate a doctrine supposedly dead: the doctrine of unexperienceable, unknowable "Things in themselves" (MW6, 4).

Dewey's rejection of things-in-themselves is so evident that Reichenbach himself points out this aspect of his thought in the essay in the Schilpp volume. Analyzing the reasons that he believes to lie at the basis of Dewey's rejection of a realistic interpretation of scientific objects, Reichenbach states that "Dewey wants to avoid this 'duplication' of things" – namely, the ontological commitment to the existence of scientific objects – because he is "afraid that it might lead to the conception of transcendent things such as Kant's 'things-in-themselves'" (Reichenbach 1939, 168). Reichenbach is wrong in establishing a connection between Dewey's rejection of things-in-themselves and the latter's (alleged) scientific antirealism; nonetheless, he is certainly right in highlighting one of the fundamental metaphysical assumptions of Dewey's philosophy of science.

Consequently, the functional distinction to which Godfrey-Smith calls attention is backed by a strong metaphysical commitment on the part of Dewey to the denial of any form of unexperienceable reality. As will be shown in the following section, this constraint has a lot to do with the kind of scientific realism that Dewey is legitimate to endorse.

What about OSR? Godfrey-Smith is quite radical in arguing that Dewey's realism cannot be of this stripe. He is right, I think, but I feel that things are a little bit more complicated than he seems to believe. First of all, let's see how Godfrey-Smith deals with this issue.

His argument is composed of two main different claims. On the one hand, he remarks that, seen from a Deweyan point of view, OSR would entail an over-intellectualization of experience. Godfrey-Smith states this point in the following manner: "the ontic version is an example of a view holding that if the cognitive side of our lives (exemplified by science) has no concern with some putative kind of entity, that is reason to think it does not exist at all" (Godfrey-Smith 2011, 78). Now, it is beyond doubt that OSR has a strong eliminativist metaphysical component which Dewey's philosophy of science does not share.[10] As Godfrey-Smith puts it, "[t]he cognitive side of our lives is concerned with relations", but, though extremely relevant, from a Deweyan perspective it does not exhaust the totality of human experience: "there is also the noncognitive side" which cannot be boiled down to the cognitive side (Godfrey-Smith 2011, 78).

On the other hand, Godfrey-Smith correctly points out that Dewey is critical of the idea that there could be a relation without some entities that the relation relates. This is OSR's distinctive thesis; so, its rejection amounts *ipso facto* to the rejection of the whole approach. Now, in *Experience and Nature* Dewey explicitly endorses the view that there cannot be relations without relata. Against the idea that structure has some sort of "superlative reality" – namely, a reality that cannot be explained and accounted for in empirical terms – he states that "all structure is structure of

[10]For the sake of completeness, it has to be acknowledged that there are also non-eliminativist versions of OSR. For this reason, it is not properly correct to argue that OSR is necessarily committed to metaphysical eliminativism. According to this moderate form of OSR, there are relations and relata, but relata are nothing more than the relations in which they stand. A non-eliminativist form of OSR is articulated in Esfeld (2004).

something", and, more precisely, that "anything defined as structure is a character of *events*, not something intrinsic and *per se*" (LW1, 64). Even in this case, Godfrey-Smith traces the reason of Dewey's disagreement with OSR to the latter's tendency to treat the cognitive access to phenomena as the sole possible access available to human beings. In his own words, OSR is the outcome of a conceptual confusion. It stems from "combining the insight that scientific theories are concerned to represent relations with the mistaken tendency to think that a scientific inventory of the world is *the* inventory of the world" (Godfrey-Smith 2011, 78).

All this said, two points have nonetheless to be stressed. First of all, Dewey explicitly and repeatedly states that scientific object *is* the relation between a set of changes experimentally brought about by an inquirer and the set of consequences that those changes cause. It is not immediate to translate that Deweyan thesis into the language of contemporary structural realism, but it is evident that Dewey does *not* believe that there are some entities, unknowable or unknown, which constitute the relata in which the relations that constitute scientific objects are grounded. In this sense, Dewey's insistence on relations has some points of contact with OSR. But here is the second point that is worth emphasizing. The relata to which Dewey makes reference are not postulated or unobservable entities, but rather the empirically accessible data that scientists produce in their experimental activities.

Recall what Dewey writes in *Experience and Nature* about the relation between primary and secondary experience. In that context, Dewey resorts to the metaphor of the vine of pendant theory which is attached at both ends to the pillars of observed subject-matter (LW1, 11; see above Sect. 1.5). So, the qualities of Dewey's relata are not the intrinsic properties which contemporary structural realists have in mind when they tackle the issue of the nature of those relata, with all the problems that the latter implies – quidditism, Ramseyan humility, and so on.

Godfrey-Smith lucidly acknowledges this point.[11] However, if I understand him correctly, the consequences that he draws from it go quite in the opposite direction, and lead to some misleading conclusions. Godfrey-Smith correctly holds that Dewey has nothing to say about the intrinsic qualities of unobservable things. And yet he concludes that those intrinsic qualities nonetheless exist, and that Dewey's philosophy does not have the theoretical resources to account for them. His argument runs as follows: (1) the qualities of which Dewey speaks are the qualities of 'middle-sized', ordinary objects; (2) this leaves the possibility open that there might be other qualities to which we do not have access in our primary experience; (3) there are powerful arguments against the theoretical coherency of OSR, since it seems contradictory to hold that there are relations without relata;

[11] Here is how Godfrey-Smith formulates the difference between Dewey's approach and structural realism: "Dewey says that the qualitative is not remote, because we have dealings with these features of the world, though they are not cognitive dealings. But these direct and non-cognitive interactions would only seem to connect us with the qualitative features of what are sometimes called "middle-sized" objects (which Dewey calls objects of "primary experience"), rather than the intrinsic qualities of aspects of the world that are extremely small, or otherwise removed from ordinary experience" (Godfrey-Smith 2011, 79).

(4) consequently, Dewey's philosophy of science, together with his theory of experience, is explanatory insufficient.

But this is clearly not what Dewey has in mind. Dewey would rather deny that there are other entities beyond the relations that *are* the scientific objects. It is for this reason that I have said above that Godfrey-Smith's dismissal of the assimilation of Dewey's scientific realism with OSR is a little bit too rash. Clearly, the point is not to argue that Dewey would agree with contemporary upholders of OSR on their most controversial theses, but rather that he shares their rejection of the very idea of the existence of intrinsic qualities on the ground that knowledge is not concerned with the apprehension of the qualities of things. In other words, as far as OSR's farewell to objects is viewed as a thesis on the nature of knowledge rather than as a metaphysical position, it can be said to pursue some of the issues that are held deer by Dewey himself. But such shift is not as innocent and unproblematic as it may look at first sight: rather the contrary, it entails a radical transformation of the philosophical framework within which to articulate the notion of structure.

5.4 Articulative Realism

After having discussed the structural realist interpretation of Dewey's scientific realism, let's go back to Dewey's reply to Reichenbach. As said, the problem with which Dewey is concerned is that of finding a possible way to combine in a consistent way his empiricism with *some* of the demands of scientific realism. Reichenbach believes that such combination is not theoretically consistent, and that Dewey's empiricism goes necessarily hand in hand with an instrumentalist and anti-realist interpretation of scientific objects. Dewey not only rejects that conclusion, but also challenges the entire argument put forth by Reichenbach. The concept of sets of relations and connections is the notion that enables Dewey to achieve the needed synthesis between empiricism and scientific realism.

Before starting the analysis, however, a conceptual consideration is in order. As has been noticed above, it is not difficult to find passages in Dewey's works in which he uses the notion of structure; nonetheless, it has also to be noticed, for the sake of honesty, that he usually prefers to speak of connections or relations between sets of observable changes. Dewey's reply to Reichenbach is a clear example of his terminological choice: there is only one occurrence of the term 'structure' in the whole text, and it is irrelevant for the issue we are concerned with here – the term indeed appears in the expression "atomic structure" (LW14, 23). Now, such difference is philosophically relevant. In contemporary philosophy of science the notion of structure is understood in logical terms, as a higher-order formal property of relations. Dewey's notion of structure is not as theoretically refined. Take, for instance, the passage from *Experience and Nature* that has been quoted above, in which Dewey states that all structure is, necessarily, structure of something: it is patent that what he has in mind in that context is the idea of structure as "constancy of means" or things used for bringing about certain consequences (LW1, 64). This loose conception of structure cannot be equated to the highly-refined notion of structure, understood as a set-theoretical constructs of the form $\langle A, R \rangle_{i \in I}$, where A is a non-empty set and R_i is a sequence of relations on A. For this reason, in this section I prefer to completely drop the term structure altogether, and rather speak of connections and relations. In so doing, I hope to avoid any possible confusion that may stem from a too quick assimilation of two different notions.

In Chap. 2 I have analyzed at length Dewey's philosophy of language. The main goal of that chapter was to introduce the distinction between significance and meaning, for the purpose of setting the ground for the semantic interpretation of the notion of inquiry that I have laid out in Chap. 4. In that chapter, it was highlighted that the couple significance/meaning goes hand in hand with other relevant conceptual distinctions. One of them is that between *connection* and *relation*. That distinction is formulated in the following terms:

> I shall reserve the word *relation* to designate the kind of "relation" which symbol-meanings bear to one another as symbol-meanings. I shall use the term *reference* to designate the kind of relation they sustain to existence; and the words *connection* (and *involvement*) to designate that kind of relation sustained by things to one another in virtue of which *inference* is possible (LW12, 61).

Accordingly, in Dewey's terminology relation and connection are by no means synonymous. 'Connection' refers to the existential relation that holds between different *things*, while 'relation' is the name Dewey gives to the relation between *symbols*. True, the distinction between existential connection and linguistic relation is not as sharp and clear-cut as it may look at first glance, but the two kinds of relations are clearly distinguished in their scope and function (see above Sect. 2.3).

With this in mind, it is now possible to formulate Dewey's scientific realism in a more appropriate manner. Recall Reichenbach's criticism that Dewey's identification of scientific objects with *relations* leads to an instrumentalist and anti-realist philosophy of science. We have seen above that Dewey rejects Reichenbach's criticism on the ground that the latter depends on what the former calls the traditional particularistic empiricism, to the effect that – to quote again his own words – "'relations' have not the empirical reality possessed by things and qualities, so that attribution of the same view to me [Dewey] logically makes relational objects unreal" (LW14, 20).

In light of the distinction between connections and relations, Dewey's reply to Reichenbach acquires a definite and precise form. In reality, Dewey does *not* say that *relations* have the empirical reality possessed by things and qualities: though the conceptual content or meaning of symbols can be – and normally is, apart from the case of inquiries – identical to the content or significance of natural signs, the relations between the former still hold exclusively on a linguistic level. What is real – as real as things and qualities are – are the connections that put things in 'relation' one to the others.

Dewey's ontology is, therefore, composed of things, qualities, and connections. *Ordinary* experience is pervaded by connections among things, and these connections ultimately depend on the *general* ways of behavior that constitute the biological and cultural endowment of the agent. The latter thesis is a corollary of Dewey's view of experience as a double-barrelled notion, which encompasses the subject-matter of life-behaviors as well as their specific ways of dealing with it. As Dewey puts it, "if one starts with the biological-cultural approach to the theory of experiencing" like the one that he develops in the first chapters of the *Logic*, "the presence of native and acquired (like habits) *general* ways of behavior is an unescapable datum" (LW14, 20–21). Now, it is worth noting that the fact that connections are directly experienced in our ordinary, common-sense life-behaviors does not immediately entail the *reality* of linguistic relations. This is another relevant difference between Dewey's philosophy of science and contemporary structural realism. Relations are the linguistic *formulations* of connections found "in the subject-matter of direct experience", but are not, properly speaking, real in the same sense, and to the same extent, in which the connections that they formulate are (LW14, 20).

I will get back to this point in a moment, since it is central to understanding what kind of scientific *realism* Dewey's is. For now I would like to focus on the idea of formulation, around which Dewey's reply to Reichenbach revolves. The notion of formulation is introduced by Dewey to account for the relation between scientific objects and qualitative experience. In this sense, it is far from surprising that Reichenbach reads Dewey's philosophy of science as a form of instrumentalism. Indeed, without a satisfactory account of its nature and function, it is rather natural to understand Dewey's notion of formulation – and accordingly, his realism about scientific objects – in instrumentalist terms.

One rather natural way of understanding the idea of formulation is the following: some connections between different features of ordinary, every-day objects are noted – for instance, the connection between the changes in the density of media and the changes in the refractive index of light – and, subsequently, are formulated through and by linguistic means. In so doing, linguistic relations formulate, on a symbolic level, existential connections among things. Consequently, scientific objects can be semantically reduced, through the medium of language, to the correlations existing between observable things. However, this view amounts to a paradigmatic form or instrumentalism. Accordingly, this way of framing the notion of formulation cannot be correct.

We are thus led back again to Dewey's philosophy of language, and, in particular, to his idea of the articulative power of linguistic symbols. In Chap. 2 I have compared Dewey's pragmatist theory of concepts with Bridgman's instrumentalist account: the goal of that analysis was to show that, all the similarities notwithstanding, the two approaches are remarkably different in that, contrary to the latter, the former does not identify the meaning of a scientific concept with a set of *fixed* operations. In so doing, the pragmatist account is capable of explaining the process of modification, development and evolution that our concepts undergo in inquiry. I have called this feature of Dewey's semantic holism 'articulation', and I have argued that articulation should be understood as a process through which new meanings are created, which accrue on the already existing ones.

We are now confronted with the very same distinction, transported from a semantic to an ontological plane. My proposal is to understand the concept of formulation – to which Dewey heavily resorts in his reply to Reichenbach, but which he uses with the same meaning in other texts as, for instance, in *The Quest for Certainty* – as an articulative notion. It is for this reason that I call Dewey's scientific realism articulative realism: according to my reading, scientific objects are the *articulative* formulation of the existential connections directly experienced in our primary transactions with the world. Such formulation is articulative in that it does not consist in the mere statement of the connections already established between a set of acts by the inquirer and a set of responses of the objects within an experimental context. Rather the contrary, it introduces a dynamic element which is absent in the instrumentalist account: the linguistic formulations of existential connections open new possibilities of action, as a consequence of which new kinds of objects are established.

Dewey's articulative realism is integral with his pragmatist theory of concepts. It should be evident, therefore, why and on what basis Dewey rejects Reichenbach's instrumentalist reading of his philosophy of science. Nonetheless, one may still raise a legitimate doubt about whether this particular stripe of realism should count as a genuine form of *realism*. In the last analysis, there seem to be good reasons to resist such claim: scientific realism seems to be essentially committed to the ontological thesis that there are mind-independent objects out there, to which the theoretical terms of our best scientific theories refer. True, structural realism – and, in particular, OSR – is committed to the existence of structures rather than objects; it is a realism about the modal structure of our world rather than about its alleged ultimate components. In any case, scientific realism – in any of its possible forms – is committed to a strong metaphysical belief in the existence of the entities postulated by our best scientific theories. This belief seems at odds with Dewey's insistence on the linguistic nature of scientific objects.

I think that a satisfactory response to that challenge has to address two different and rather independent problems: the first issue concerns the criteria of objectivity that have to be satisfied in order for a philosophy of science to count as genuinely realist; the second issue has to do with the idea that linguistic relations are less real – in some intuitive sense – than the existential connections that they purport to formulate. Within Dewey's framework, those two issues are closely connected, but it is useful to treat them separately.

Let's start with the first issue. As has been highlighted in Chap. 4, according to Dewey's functional conception of objectivity, an object is anything which has inferential stability. Objectivity is to be cashed out exclusively in terms of the inferential connections that can be drawn from the existence of a certain quality or element. Now, a remarkable difference has to be acknowledged between the inferential stability that lies at the basis of common-sense objects and the one that constitutes to scientific objects. In the first case, the scope and depth of the inferences that can be drawn from the actual presence of a certain element in the field of primary experience is extremely limited: for instance, the transparency of water is a reliable sign of its drinkability and of its capacity to wash hands. In the case of scientific objects, the qualities that are so prominent in our primary experience – and which we normally use as a solid ground of inference – lose their evidential import, and, as a consequence of this fact, the kind of objects that are thus constructed are purely and exclusively relational.

This said, the reality of scientific objects is structurally and functionally identical to the reality that we attribute to common-sense objects, and the property of being an object is attributed to the latter on the very same basis on which it is attributed to the former. Accordingly, at least from Dewey's perspective, there is no possible room to question the reality – in the sense of objectivity – of scientific objects by contrasting them with ordinary objects. The difference in the inferential depth and scope which tells ordinary objects from scientific entities presupposes a common conception of the nature of objectivity.

Dewey is explicit on this point. In *The Quest of Certainty* he formulates his continuist account of objectivity with the following words:

> As habits form, action is stereotyped into a fairly constant series of acts having a common end in view; *the* table serves a single use, in spite of individual variations. A group of properties is set aside, corresponding to the abiding end and single mode of use which form *the* object, in distinction from "this" of unique experiences. *The* object is an abstraction, but unless it is hypostatized it is not a vicious abstraction. It designates selected relations of things which, with respect to their mode of operation, are constant within the limits practically important [...]. The scientific or physical object marks an extension of the same sort of operation. *The* table, as *not* a table but as a swarm of molecules in motions of specified velocities and accelerations, corresponds to a liberated generalization of the purposes which *the* object may serve. "Table" signifies a definite but restricted set of uses; stated in the physical terms of science it is thought of in a wider environment and free from any specified set of uses; out of relation to any particular individualized experience. The abstraction is as legitimate as is that which gives rise to the idea of the table, for it consists of standardized relations or interactions. (LW4, 190).

This passage not only makes it clear why Dewey holds that there is no *specific* problem with the commitment to the reality of scientific objects; it also allows to understand the seemingly elusive solution that Dewey gives to the so-called Eddington's two-tables paradox. As has been noticed in Chap. 2, Dewey was strongly influenced by the operationalist tradition, to such an extent that, with some degree of exaggeration, *The Quest for Certainty* can be said to originate from his critical assimilation of Bridgman and Eddington's works. Now, one of the passages that impressed Dewey the most is certainly the opening remarks of Eddington's *The Nature of the Physical World*. In those pages, Eddington contrasts two different tables. One is the 'commonplace object' of the world, the table with which anyone is familiar from her early days: this table "has extension; it is comparatively permanent; it is coloured; above all it is substantial" (Eddington 1928, ix). The other table is the scientific table, which is not part of the commonplace world. There is nothing substantial, Eddington notices, in this table: it is "nearly all empty space", and even "in the minute part which is not empty" it would be better to refrain from using the familiar category of substance. As Eddington puts it, "[i]n dissecting matter into electric charges we have traveled far from that picture of it which first gave rise to the conception of substance", as a consequence of which "the meaning of that conception – if it ever had any – has been lost by the way" (Eddington 1928, x-xi).

The contrast between the two tables is intended to call into question our intuitions about the notions of reality and objectivity: it asks to take a position on which kind of entities one should be ontologically committed to – or, stated in other terms, whether one should acknowledge ontological primacy to the scientific image or to the manifest image of the world. As is well-known, Eddington's solution to the two-tables problem that bears his name is radical: he believes that only the scientific table is really there, whereas the commonplace table is an illusion. Dewey cannot agree with Eddington. His theory of objectivity forces him to the conclusion that the two types of objects are equally real – so to say – in that they satisfactorily perform the function that they are constructed for. Scientific and ordinary objects are equally real in their respective contexts of use.

In *The Quest for Certainty*, such an irenic ontological attitude is formulated as follows:

> Water as an object of science, as H_2O with all the other scientific propositions which can be made about it, is not a rival for position in real being with the water we see and use. It is, because of experimental operations, an added instrumentality of multiplied controls and uses of the real things of everyday experience (LW4, 85).

As should now be clear, this approach does not count, and should not be read, as a form of instrumentalism. Dewey's definition of scientific object as an "added instrumentality of multiplied controls and uses of the real things of everyday experience" is to be understood within the framework of his pluralistic and anti-reductionist account of experience. The rivalry between scientific and commonsense objects is thus defused by rejecting the whole theoretical scenario in which the very possibility of such rivalry is set up.[12]

[12]Because of the importance of this problem, it comes as no surprise that Dewey deals with it at some length in his reply to Reichenbach. The passage immediately quoted, indeed, may be well read in instrumentalist terms: this is the kind of reading that Reichenbach advances and the one that Dewey wants to criticize and reject. The instrumentalist reading seems to be corroborated by another passage on *The Quest for Certainty*. The passage runs as follows: "[t]he problem which is supposed to exist between two tables, one that of direct perception and use and the other that of physics (to take the favorite illustration of recent discussion) is thus illusory. The perceived and used table is the only table, for it alone has both individuality of form – without which nothing can exist or be perceived – and also includes within itself a continuum of relations or interactions brought to a focus" (LW4, 191). Now, in his reply to Reichenbach Dewey reformulates these ideas in slightly different terms. Directly referring to the passage from *The Quest for Certainty*, he writes: "'The perceived and used table is the only table'. Now this passage, because of the use of the word *only*, might be taken to deny that a scientific physical object exists. If the passage had read: 'The perceived and used table is the only *table*', the italics might have warded off misinterpretation. For it would have indicated that it was not the existence of a swarm of atoms (electrons, etc.) in rapid movement which was denied, but the notion that this swarm somehow constitutes a ghostly kind of *table*, instead of being just what it is in terms of electrons, deuterons, etc. One would hardly put books or dishes on the latter or sit down before it to eat. That the table as a perceived table is an object of knowledge in one context as truly as the physical atoms, molecules, etc., are in another situational context and with reference to another *problem* is a position I have given considerable

We are thus led to the second issue under discussion, namely, the allegedly anti-realistic implications of the idea that linguistic relations are less real than the existential connections that they formulate. Intuitively, it seems plausible to hold that a realistic standpoint should be committed to the view that scientific objects are *connections* rather than relations: in the last analysis, connections are existential features of the world, whereas relations hold on a linguistic plane. So, one may be led to ask, why not shift attention from relations to connections, so as to secure the realist component of Dewey's articulative realism?

From Dewey's perspective, there are sound theoretical reasons for resisting the identification of scientific objects with existential connections. In particular, it is relevant to note that, contrary to what one might be ready to believe, far from leading to a full-fledged realism, such an identification would issue in a form of instrumentalism. In order to realize this point, it is useful to consider the following passage from Dewey's reply to Reichenbach:

> the definitely relational objects of science are produced when *connections* existing in the immediate situation are noted and formulated, the latter process involving elaboration in discourse. If, however, the operational presence of *general* modes of activity (constituting *connections*) in the material of ordinary experience is ignored (if, in the sense in which "thing" is an equivalent of the Latin *res*, it is not noted that a way of behaving is a "thing"), then the general and relational character of scientific objects must be denied by a professed empiricist (LW14, 23).

Since scientific objects, whose nature is explicitly said to be *relational*, are constructed when existing connections are noted and formulated, they cannot be reduced to the latter. As previously remarked, the articulative function of language is a necessary component of Dewey's realism. Now, it is such an articulative capacity of language which makes it possible to attribute reality to *scientific* objects.

Indeed, the connections to which Dewey refers as the raw material that linguistic relations formulate and articulate are grounded in the "general modes of activity in the material of *ordinary* experience" (italics mine). Accordingly, they are nothing but the existential connections that are produced and brought about within experimental contexts of activity: for instance, I change the density of media and I register that a change in the refractive index of light occurs. These existential connections are given in primary experience. Considered as *connections*, they are unproblematic, and are naturally taken to be part of the furniture of the world. At this level, however, no scientific object is given. In order for a scientific object to exist, primary experience has to give way to secondary experience – that is, to a series of highly refined activities whose exclusive concern is that of "inquiry *as inquiry*" (LW16, 254). In other words, in order for a scientific object to arise, existential connections have to be translated and interpreted into the terms of a certain scientific theory or framework. It is only as a consequence of a process of linguistic articulation that the meaning of these existential connections is enriched,

space to developing" (LW14, 22). In so doing, by reformulating the distinction between the two tables in terms of the different life-behaviors in which they are used, instrumentalism loses its bite.

and the latter become part of a broader web of relations which have explanatory power.[13]

So, it is the articulative function of language that makes Dewey's articulative realism a full-fledged *realism*. Without the process of articulation that transforms existential connections into linguistic relations, the acknowledgment of the existence of connections in the material of ordinary experience would issue into a form of instrumentalism. By stressing the indispensability of linguistic articulation, on the contrary, Dewey denies that connections alone are sufficient to constitute scientific objects. Which means, on the one hand, that different scientific objects are compatible with the same set of existential connections; and, on the other hand, that the attribution of reality to scientific objects depends on the availability of a theoretical framework in which existential connections are thought and understood.

The latter remark introduces my final point about the relationship between relations and connections. In the passage quoted above, Dewey recognizes a certain degree of asymmetry between relations and connections. As repeatedly noticed, connections among ordinary things are 'there', independently of the existence of a scientific theory or framework that provides an explanation of their features and which, in so doing, constructs a scientific object that performs that explanatory task. On the contrary, the reality of relations – and, consequently, of scientific objects – is dependent upon the reality of connections. I think this conclusion can be safely drawn from Dewey's statement that if "the operational presence of general modes of activity [...] in the material of ordinary experience is ignored", then one must deny "the general and relational character of scientific objects" (LW14, 23).

Connections 'underdetermine' relations. Taken in itself, this form of underdetermination is clearly unproblematic. Equally unproblematic is, I think, the assumption that connections are more real than relations, provided that this thesis is taken as consisting in the claim that connections are far less amenable to revision than the scientific objects that are constructed for the purpose of accounting for them. This said, the asymmetry between connections and relations is used by Dewey to develop an account of what it takes for a set of existential connections to become a scientific object.

The account runs as follows:

> Suppose one of those persons of extraordinarily keen vision who abound in the Grimm fairy tales were in fact to see, sensibly to perceive, an object which had all the qualities a physicist attributes to the atom. He would surely see something. But would he see an atom in the definite sense of seeing that which is *an object of physical science*? I can find but one possible answer, namely: "It depends. If he himself has had a scientific training and if in sensibly perceiving this particular thing he explicitly *identifies* it as having all the *relational* properties required by the scientific theory of atomic structure and with no properties incompatible with the latter, the answer is Yes. But if he sees it merely as another man of lesser power of vision sees a rock, the answer is No." In other words, it is not just the thing as perceived, but the thing as and when it is placed in an extensive ideational or

[13]Recall here what Dewey writes about the process of meaning liberation in *Logic* (LW12, 58). On this point, see below Chapter 2.5.

theoretical context *within which it exercises a special office* that constitutes a distinctively physical scientific object (LW14, 24).

Dewey's example is extremely interesting. It formulates the difference existing between primary and secondary experience in terms of the actions that the agent is capable of doing. Dewey's claim is that if the agent is capable of responding to the features of commonsense objects in terms of the relations stated by the theory, then she can be said to *see* the corresponding scientific object. Otherwise, if she has not access to the scientific theory, or if she is not capable of using the meanings of scientific concepts to enrich the significances of the observed things, then she sees a commonsense object.[14]

---

[14]In the article *Dewey and Quine on the Logic of What There Is*, Shook has correctly pointed out that one of the functions that Dewey attributes to scientific theories is that of enabling the agent to intellectually see an observable thing in a radically new way: "the concept of a 'voltage meter'", he writes, "is a theoretical means of 'seeing' an object in a way unavailable to those not yet initiated into the theory of electrons (Shook 2002, 106). I would like to call attention to the conclusion that Shook draws from that remark. He notices that this function of scientific theories "permits discussion of a *transformed* object created by a theory, existing within the experience of those trained in the theory". Secondary experience is the locus in which these highly refined entities exist. "The transformed entity", Shook concludes, "is an independent but not transcendent entity, since it can be experienced, yet it is experienced as the sort of thing that exists whether or not it is experienced" (Shook 2002, 106). I wholly agree with Shook on this point. I do not agree with him, on the contrary, when he contrasts transformed objects with transcendent postulated entities. I quote the passage in which he formulates such contrast in its entirety. "Transcendent postulated entities have four characteristics very different from transformed objects according to Dewey's functionalist epistemology: They can never be observed; they are only hypothetical entities; their meaning as hypothetical is exhausted by their definitions; and they are imaginative creations of theories that are not only fallible but underdetermined by any amount of practical success. Transcendent postulated entities have these characteristics because their meaning consists entirely of their role in 'universal' and 'hypothetical' propositions of scientific law" (Shook 2002, 107). If I understand it correctly, Shook's proposal runs as follows: (1) transcendent postulated entities are introduced and defined in scientific laws; (2) within Dewey's framework, scientific laws concerning this kind of entities are universal or hypothetical propositions; (3) universal propositions do not make existential claims, at least in the first place, since they simply provide the definition of the concepts; (4) so, we are not ontologically committed to the existence of transcendent postulated entities. For the sake of honesty, it has to be acknowledged that Shook's argument is much more articulated than how I have presented it; nonetheless, I think that those four claims faithfully summarize its relevant content. The reason why I disagree with Shook is that I do not see any sound reason to distinguish between transformed objects and transcendent postulated entities. According to the semantic reading of inquiry that I have laid out in Chap. 4, universal propositions are not truth-bearers. Consequently, it is correct to say – as Shook does – that we are not ontologically committed to the existence of transcendent postulated entities that are formulated in universal propositions. However, propositions are means whose function is to contribute to the reconstruction of the problematic situation that called out inquiry. Now, when the predicate of the judgment – in which universal propositions eventually converge – is existentially predicated of the subject, the meanings contained in the predicate are attributed to the significances of the object. The ways in which the agent transacts with the object are now mediated by the concepts of the scientific theory. At *this* level, we are ontologically committed to the existence of the transcendent postulated entities that are formulated in universal propositions. Contrary to Shook, I reject the

Obviously, that claim is in strong agreement with Dewey's functional theory of objectivity, as well as with his linguistic account of experience and inquiry. Nonetheless, its philosophical import can be questioned. What does Dewey mean to say with that example? Is he merely trying to formulate the epistemic conditions under which an agent can be said to know that she is dealing with a scientific object? Or is he stating a much stronger, ontological thesis, to the effect that a scientific object is constituted only when an agent is capable of treating it as such?

I am inclined toward the latter view. More importantly, it is Dewey himself who argues for it: by saying that it is only when the thing "is placed in an extensive ideational or theoretical context *within which it exercises a special office* that constitutes a distinctively physical scientific object", he gives an ontological twist to the entire argument (LW14, 23). This quotation confirms, therefore, that, according to Dewey's articulative realism, scientific objects are constituted *as objects* within secondary experience, and that they are made possible by a semantic process of translation of existential connections into the terms of a scientific theory. As Dewey points out, a distinctively physical scientific object is constituted when the perceived thing – the object that is constituted in our common-sense transactions with the environment – is identified as having all the relational properties that a certain scientific theory predicates of it. The object is thus made as one of a *kind*, and it is acted upon accordingly.

As is evident, Dewey's scientific realism is not metaphysical: it does not assume the existence of scientific objects – or, better said, of the objects that are the referents of scientific concepts – as a brute ontological fact. This should come as no surprise. Indeed, the metaphilosophical primacy that Dewey acknowledges to the category of activity puts constraints on what kind of realism can be legitimately held as theoretically consistent. And, clearly, metaphysical realism does not satisfy the requirements of Dewey's metaphilosophical approach.

Consequently, Dewey's scientific realism cannot help but be a *practical* realism, its distinctive feature being the idea that ontological commitment is *practice-sensitive*. For something to be a scientific object means that the agent has the capacity to take part into scientific life-behaviors, and to undertake some specific activities that make it possible for scientific potentialities to accrue on the significances of common-sense objects. The reality of scientific objects is thus traced back to the highly complex structure of scientific life-behaviors. Scientific ontology is the linguistic inventory – recall that scientific objects are *statements* in a scientific language – of the ways in which we deal with the objects of the world while we are exclusively concerned with the advancement of knowing. What we *can* do with an object establishes what there is. Dewey's philosophy of science, in all its different components (semantic, epistemic, ontological), can be seen, therefore, as stemming from one single assumption: activity all the way down.

---

distinction, and I would rather say that transcendent postulated entities are transformed objects in the making.

References

Agazzi, E., & Heinzmann, G. (Eds.). (2015). *The practical turn in philosophy of science*. Milano: Franco Angeli.

Aikin, S. F., & Talisse, R. B. (2017). *Pragmatism, pluralism, and the nature of philosophy*. New York/London: Taylor & Francis Ltd.

Alexander, T. (2018). Dewey's naturalistic metaphysics. In S. Fesmire (Ed.), *The Oxford handbook of Dewey*. Oxford: Oxford University Press.

Alexander, T. M. (1987). *John Dewey's theory of art, experience, and nature: The horizons of feeling*. Albany: SUNY Press.

Alexander, T. M. (2004). Dewey's denotative-empirical method: A thread through the labyrinth. *The Journal of Speculative Philosophy, 18*(3), 248–256.

Almeder, R. (2007). Pragmatism and philosophy of science: A critical survey. *International Studies in the Philosophy of Science, 21*(2), 171–195.

Anderson, D. R. (2006). *Philosophy Americana*. New York: Fordham University Press.

Barrotta, P. (2018). *Scientists, democracy and society. A community of inquirers*. Cham: Springer.

Barrotta, P., & Gronda, R. (2019). Scientific experts and citizens' trust: Where the third wave of social studies of science goes wrong. *Teoria*, 9–27.

Bird, A. (2007a). *Nature's metaphysics: Laws and properties*. Oxford: Oxford University Press.

Bird, A. (2007b). The regress of pure powers? *The Philosophical Quarterly, 57*(229), 513–534.

Black, M. (1962). Dewey's philosophy of language. *The Journal of Philosophy, 59*(19), 505–523.

Boersema, D. (2008). *Pragmatism and reference*. Cambridge, MA: MIT Press.

Boisvert, R. D. (1988). *Dewey's metaphysics*. New York: Fordham University Press.

Brading, K., & Crull, E. (2017). Epistemic structural realism and Poincaré's philosophy of science. *HOPOS: The Journal of the International Society for the History of Philosophy of Science, 7(1)*, 108–129.

Brandom, R. (2000). *Articulating reasons*. Cambridge, MA: Harvard University Press.

Brandom, R. (2007). Inferentialism and some of its challenges. *Philosophy and Phenomenological Research, 74*(3), 651–676.

Bridgman, P. W. (1927), *The logic of modern physics*. New York: The Macmillan Company.

Brodbeck, M. (1949). The new rationalism: Dewey's theory of induction. *The Journal of Philosophy, 46*(24), 780–791.

Brown, M. J. (2009). Models and perspectives on stage: Remarks on Giere's scientific perspectivism. *Studies in History and Philosophy of Science Part A, 40*(2), 213–220.

Brown, M. J. (2012). John Dewey's logic of science. *HOPOS: The Journal of the International Society for the History of Philosophy of Science, 2*(2), 258–306.

R. Gronda, *Dewey's Philosophy of Science*, Synthese Library 421,
https://doi.org/10.1007/978-3-030-37562-1

Browning, D. (1998). Dewey and Ortega on the starting point. *Transactions of the Charles S. Peirce Society, 34*(1), 69–92.

Browning, D. (2002). Designation, characterization, and theory in Dewey's logic. In *Dewey's logical theory. New studies and interpretations* (pp. 160–179). Nashville: Vanderbilt University Press.

Burke, T. (1994). *Dewey's new logic. A reply to Russell.* Chicago: The University of Chicago Press.

Burke, T. (2000). What is a situation? *History and Philosophy of Logic, 21*(2), 95–113.

Burke, T. (2002). Qualities, universals, kinds, and the new riddle of induction. In F. T. Burke, D. M. Hester, & R. B. Talisse (Eds.), *Dewey's logical theory. New studies and interpretations* (pp. 225–235). Nashville: Vanderbilt University Press.

Burke, T. (2009). Browning on inquiry into inquiry, part 2. *Transactions of the Charles S. Peirce Society, 45*(2), 157–176.

Campbell, J. (1995). *Understanding John Dewey: Nature and cooperative intelligence.* La Salle: Open Court.

Capps, J. (1996). Dewey, Quine, and pragmatic naturalized epistemology. *Transactions of the Charles S. Peirce Society, 32*(4), 634–667.

Capps, J. (2017). A pragmatic argument for a pragmatic theory of truth. *Contemporary Pragmatism, 14*(2), 135–156.

Capps, J. (2018). Did Dewey have a theory of truth? *Transactions of the Charles S. Peirce Society, 54*(1), 39.

Cartwright, N. (1983). *How the laws of physics lie.* Oxford: Oxford University Press.

Cartwright, N. (1999). *The dappled world.* Cambridge, MA: Cambridge University Press.

Chang, H. (2011a). The persistence of epistemic objects through scientific change. *Erkenntnis, 75*(3), 413–429.

Chang, H. (2011b). The philosophical grammar of scientific practice. *International Studies in the Philosophy of Science, 25*(3), 205–221.

Chang, H. (2012). *Is water H2O?* Dordrecht: Springer.

Chang, H. (2014). Epistemic activities and systems of practice. Units of analysis in philosophy of science after the practice turn. In L. Soler, S. Zwart, M. Lynch, & V. Israel-Jost (Eds.), *Science after the practice turn in the philosophy, history, and social studies of science* (pp. 67–79). London: Routledge.

Chang, H. (2016). Pragmatic realism. *Revista de Humanidades de Valparaíso, 4*(8), 107–122.

Coffa, J. A. (2003). *The semantic tradition from Kant to Carnap. To the Vienna station.* Cambridge, MA: Cambridge University Press.

Demopoulos, W., & Friedman, M. (1985). Bertrand Russell's "The analysis of matter": Its historical context and contemporary interest. *Philosophy of Science, 52*(4), 621–639.

Dewey, J. (2012). *Unmodern philosophy and modern philosophy.* Carbondale: Southern Illinois University Press.

Dreon, R. (2014). Dewey on language: Elements for a non-dualistic approach. *European Journal of Pragmatism and American Philosophy, VI*(2), 109–124.

Dupré, J. (1995). *The disorder of things.* Cambridge, MA: Harvard University Press.

Eames, S. M. (1964). Primary experience in the philosophy of John Dewey. *Monist, 48*(3), 407–418.

Eddington, A. (1928). *The nature of the physical world.* New York: The Macmillan Company.

El-Hani, C. N., & Pihlström, S. (2002). Emergence theories and pragmatic realism. *Essays in Philosophy, 3*(2), Article 3.

Esfeld, M. (2004). Quantum entanglement and a metaphysics of relations. *Studies in History and Philosophy of Science Part B: Studies in History and Philosophy of Modern Physics, 35*(4), 601–617.

Fesmire, S. (2014). *Dewey.* New York: Routledge.

Fine, A. (2007). Relativism, pragmatism, and the practice of science. In C. Misak (Ed.), *New pragmatists* (pp. 50–67). Oxford: Oxford University Press.

Flemming, H., Wingender, J., Szewzyk, U., Steinberg, P., Rice, S. A., & Kjelleberg, S. (2016). Biofilms: An emergent form of bacterial life. *Nature Reviews Microbiology, 14*(9), 563–575.
Flemming, H.-C. (2016). Eps – Then and now. *Microorganisms, 4*(4), 41.
Frega (2006). *Pensée, expérience, pratique. Essai sur la théorie du jugement de John Dewey*. Paris: L'Harmattan.
Frega (2010). From judgment to rationality: Dewey's epistemology of practice. *Transactions of the Charles S. Peirce Society, 46*(4), 591.
Frega, R. (Ed.). (2011). *Pragmatist epistemologies*. Lanham: Lexington Books.
Frega, R. (2012). *Practice, judgment, and the challenge of moral and political disagreement: A pragmatist account*. Lanham: Lexington Books.
French, S. (2014). *The structure of the world*. Oxford: Oxford University Press.
French, S., & Da Costa, N. C. A. (2003). *Science and partial truth: A unitary approach to models and scientific reasoning*. Oxford: Oxford University Press.
French, S., & Ladyman, J. (2003). Remodelling structural realism: Quantum physics and the metaphysics of structure. *Synthese, 136*(1), 31–56.
French, S., & Ladyman, J. (2010). In defence of ontic structural realism. In A. Bokulich & P. Bokulich (Eds.), *Scientific structuralism* (pp. 25–42). Dordrecht: Springer.
Friedman, M. (1999). *Reconsidering logical positivism*. Cambridge, MA: Cambridge University Press.
Friedman, M. (2001). *Dynamics of reason*. Stanford: Centre for the Study of Language & Information Publications.
Gale, R. (2002). The metaphysics of John Dewey. *Transactions of the Charles S. Peirce Society, 38*(4), 477–519.
Gale, R. M. (2010). The naturalism of John Dewey. In M. Cochran (Ed.), *The Cambridge companion to Dewey* (pp. 55–79). Cambridge, MA: Cambridge University Press.
Galison, P., & Stump, D. (1996). *The disunity of science*. Stanford: Stanford University Press.
Garrison, J. W. (2005). Dewey on metaphysics, meaning making, and maps. *Transactions of the Charles S. Peirce Society: A Quarterly Journal in American Philosophy, 41*(4), 818–844.
Garrison, J. W., & Shargel, E. I. (1988). Dewey and Husserl: A surprising convergence of themes. *Educational Theory, 38*(2), 239–247.
Gautier, C. (2017). Attitudes of knowledge and common sense. *European Journal of Pragmatism and American Philosophy, IX*(2), 1–23.
Giere, R. N. (1996). From wissenschaftliche philosophie to philosophy of science. In R. N. Giere & A. W. Richardson (Eds.), *Origins of logical empiricism* (pp. 335–354). Minneapolis: University of Minnesota Press.
Giere, R. N. (1999). *Science without laws*. Chicago: The University of Chicago Press.
Giere, R. N. (2006). Modest evolutionary naturalism. *Biological Theory, 1*(1), 52–60.
Godfrey-Smith, P. (2011). Dewey and the subject-matter of science. In J. Shook & P. Kurtz (Eds.), *Dewey's enduring impact: Essays on America's philosopher* (pp. 73–86). Amherst: Prometheus Books.
Godfrey-Smith, P. (2013a). Dewey and the question of realism. *Nous, 50*(1), 73–89.
Godfrey-Smith, P. (2013b). Quine and pragmatism. In G. Harman & E. Lepore (Eds.), *A companion to W.V.O. Quine* (pp. 54–68). Malden, MA; Chichester: Wiley-Blackwell.
Godfrey-Smith, P. (2019). Dewey and anti-representationalism. In S. Fesmire (Ed.), *The Oxford handbook of Dewey*. Oxford: Oxford University Press.
Good, J. A. (2006). *A search for unity in diversity: The "permanent hegelian deposit" in the philosophy of John Dewey*. Lanham: Lexington Books.
Gronda, R. (2011). Review of J. Shook & J. Good (Eds.), John Dewey's philosophy of spirit, with the 1897 lecture on Hegel. *European Journal of Pragmatism and American Philosophy, III*(2), 305–315.
Gronda, R. (2012). Sense and construction. The development of Dewey's theory of meaning (1884–1903). 2012, Ph.D. Dissertation.
Gronda, R. (2013). Review of Morse, Faith in life. John Dewey's early philosophy. *European Journal of Pragmatism and American Philosophy, V*(2), 191–197.

Gronda, R. (2015a). Normativity and objectivity. The semantic nature of objects and the potentiality of nature. *European Journal of Pragmatism and American Philosophy, VII*(1), 115–129.
Gronda, R. (2015b). Review of J. Scott Johnston, John Dewey's earlier logical theory. *European Journal of Pragmatism and American Philosophy, VII*(2), 225–231.
Gronda, R. (2015c). What does China mean for pragmatism? A philosophical interpretation of Dewey's Sojourn in China (1919–1921). *European Journal of Pragmatism and American Philosophy, VIII*(2), 45–70.
Gronda, R. (2017). Thinking under the pressure of practice: A critical interpretation of Dewey's preparatory notes for the lectures in China. *Transactions of the Charles S. Peirce Society, 53*(1), 71–89.
Gronda, R. (2018a). American context. In M. G. Festl (Ed.), *Handbuch pragmatismus* (chapter 39, pp. 290–296). Stuttgart: J. B. Metzler.
Gronda, R. (2018b). Rethinking the notion of public. A pragmatist account. In P. Barrotta & G. Scarafile (Eds.), *Science and democracy: Controversies and conflicts* (pp. 53–70). Amsterdam: John Benjamins Publishing Company.
Gronda, R., & Turbanti, G. (2017). On the pragmatic content of science and common sense. *European Journal of Pragmatism and American Philosophy, 9*(2), 1–24.
Hacking, I. (1983). *Representing and intervening. Introductory topics in the philosophy of natural science*. Cambridge, MA: Cambridge University Press.
Hacking, I. (2000). *The social construction of what?* Cambridge, MA: Harvard University Press.
Hacking, I. (2007). On not being a pragmatist: Eight reasons and a cause. In C. Misak (Ed.), *New pragmatists* (pp. 32–49). Oxford: Oxford University Press.
Hawthorne, J. (2001). Causal structuralism. *Nous, 35*(s15), 361–378.
Heil, J. (2003). *From an ontological point of view*. Oxford: Oxford University Press.
Heil, J. (2005). Dispositions. *Synthese, 144*(3), 343–356.
Hempel, C. G. (1965). The theoretician's dilemma: A study in the logic of theory construction. *Aspects of scientific explanation: Essays in the philosophy of science* (pp. 173–226). New York: The Free Press.
Hetherington, S. (2017). Knowledge as potential for action. *European Journal of Pragmatism and American Philosophy, IX*(2), 1–19.
Hickman, L. (2007). *Beyond the epistemology industry: Dewey's theory of inquiry* (chapter 12, pp. 206–230). New York: Fordham University Press.
Hildebrand, D. (2018). Dewey, Rorty, and Brandom: The challenges of linguistic neopragmatism. In S. Fesmire (Ed.), *The Oxford handbook of Dewey*. Oxford: Oxford University Press.
Hildebrand, D. L. (2003). *Beyond realism and antirealism: John Dewey and the neopragmatists*. Nashville: Vanderbilt University Press.
Howard, D. (2003). Two left turns make a right: On the curious political career of North American philosophy of science at midcentury. In *Logical empiricis in North American*. Minneapolis: University of Minnesota Press.
Ingthorsson, R. D. (2013). Properties: Qualities, powers, or both? *Dialectica, 67*(1), 55–80.
Jacobs, J. D. (2011). Powerful qualities, not pure powers. *The Monist, 94*(1), 81–102.
James, W. (1975). *Pragmatism*. Cambridge, MA: Harvard University Press.
James, W. (1976). *Essays in radical empiricism*. Cambridge, MA: Harvard University Press.
James, W. (2007). *The principles of psychology (Vol. 1)*. New York: Cosimo Inc.
Johnson, M. (2014). Experiencing language: What's missing in linguistic pragmatism? *European Journal of Pragmatism and American Philosophy, VI*(2), 14–27.
Johnston, J. S. (2014). *John Dewey's earlier logical theory*. Albany: Suny Press.
Kant, I. (1781/2016). *Critique of pure reason*. Cambridge, MA: Cambridge University Press.
Karatan, E., & Watnick, P. (2009). Signals, regulatory networks, and materials that build and break bacterial biofilms. *Microbiology and Molecular Biology Reviews, 73*(2), 310–347.
Kellert, S., Longino, H., & Waters, C. K. (Eds.). (2006). *Scientific pluralism*. Minneapolis: University of Minnesota Press.

Kestenbaum, V. (1977). *The phenomenological sense of John Dewey: Habit and meaning*. Atlantic Highlands: Humanities Press.
Kirby, C. C. (2014). Dewey and the "Greeks": Inquiry and the organic spirit of greek philosophy. In C. C. Kirby (Ed.), *Dewey and the ancients. Essays on hellenic and hellenistic themes in the philosophy of John Dewey* (pp. 47–77). London: Bloomsbury Publishing.
Kitcher, P. (2011). *Science in a democratic society*. Amherst: Prometheus Books.
Kitcher, P. (2012). *Preludes to pragmatism: Toward a reconstruction of philosophy*. Oxford: Oxford University Press.
Kitcher, P. (2014). Extending the pragmatist tradition: Replies to commentators. *Transactions of the Charles S. Peirce Society, 50*(1), 97.
Kitcher, P. (2016). Reply to Massimi. In M. Couch & J. Pfeifer (Eds.), *The philosophy of Philip Kitcher* (pp. 114–120). Oxford: Oxford University Press.
Kukla, A. (2000). *Social constructivism and the philosophy of science*. London: Routledge.
Ladyman, J. (1998). What is structural realism? *Studies in History and Philosophy of Science Part A, 29*(3), 409–424.
Hickman, L., & Reich, K. (Eds.). (2009). *John Dewey between pragmatism and constructivism*. New York: Fordham University Press.
Laudan, L. (1981). A confutation of convergent realism. *Philosophy of Science, 48*(1), 19–49.
Laudan, L. (1984). *Science and values: The aims of science and their role in scientific debate*. Berkeley: University of California Press.
Levi, I. (2010). Dewey's logic of inquiry. In M. Cochran (Ed.), *The Cambridge companion to Dewey* (pp. 80–100). Cambridge, MA: Cambridge University Press.
Levine, S. (2019). *Pragmatism, objectivity, and experience*. Cambridge, MA: Cambridge University Press.
Lewis, C. I. (1923). A pragmatic conception of the a priori. *The Journal of Philosophy, 20*(7), 169.
Lewis, C. I. (1929). *The mind and the world-order. Outline of a theory of knowledge*. New York: Charles Scribner's Sons.
Lewis, D. (1986). *Philosophical papers: Volume II*. Oxford: Oxford University Press.
Madzia, R., & Jung, M. (Eds.). (2016). *Pragmatism and embodied cognitive science: From bodily intersubjectivity to symbolic articulation*. Berlin: de Gruyter.
Margolis, J. (2002). Dewey's and Rorty's opposed pragmatisms. *Transactions of the Charles S. Peirce Society, 38*(1/2), 117–135.
Margolis, J. (2014). Richard Rorty contra Rorty and John Dewey. *European Journal of Pragmatism and American Philosophy, VI*(2), 91–108.
Martin, C. B. (1997). On the need for properties: The road to pythagoreanism and back. *Synthese, 112*(2), 193–231.
Martin, C. B., & Heil, J. (1999). The ontological turn. *Midwest Studies in Philosophy, 23*(1), 34–60.
Massimi, M. (2010a). Galileo's mathematization of nature at the crossroad between the empiricist and the kantian tradition. *Perspectives on Science, 18*(2), 152–188.
Massimi, M. (2010b). Structural realism: A neo-kantian perspective. In *Scientific structuralism* (pp. 1–23). Dordrecht: Springer.
Massimi, M. (2016). Bringing real realism back home. A perspectival slant. In M. Couch & J. Pfeifer (Eds.), *The philosophy of Philip Kitcher* (pp. 98–114). Oxford: Oxford University Press.
Massimi, M. (2018). Points of view: Kant on perspectival knowledge. *Synthese*.
Mayeroff, M. (1950). The nature of propositions in John Dewey's "logic": A reply to miss Brodbeck. *The Journal of Philosophy, 47*(12), 353–358.
Mayo, D. G. (1996). *Error & the growth of experimental knowledge*. Chicago: University of Chicago Press.
McDowell, J. (1994). *Mind and world*. Cambridge, MA: Harvard University Press.
Mesthene, E. G. (1959). The role of language in the philosophy of John Dewey. *Philosophy and Phenomenological Research, 19*(4), 511–517.
Midtgarden, T. (2008). Dewey's philosophy of language. *Revue internationale de philosophie, 245*(3), 257–272.

Misak, C. (2004). Charles Sanders Peirce (1839–1914). In C. Misak (Ed.), *The Cambridge companion to Peirce* (pp. 1–26). Cambridge, MA: Cambridge University Press.
Misak, C. J. (1995). *Verificationism. Its History and Prospects*. New York: Routledge.
Mitchell, S. (2003). *Biological complexity and integrative pluralism*. Cambridge, MA: Cambridge University Press.
Mitchell, S. (2009). *Unsimple truths. Science, complexity and policy*. Chicago: University of Chicago Press.
Molnar, G. (2003). *Powers: A study in metaphysics*. Oxford: Oxford University Press.
Morganti, M. (2004). On the preferability of epistemic structural realism. *Synthese, 142*(1), 81–107.
Morganti, M. (2011). Is there a compelling argument for ontic structural realism? *Philosophy of Science, 78*(5), 1165–1176.
Mormann, T. (2007). Carnap's logical empiricism, values, and American pragmatism. *Journal for General Philosophy of Science, 38*(1), 127–146.
Mormann, T. (2012). A place for pragmatism in the dynamics of reason? *Studies in History and Philosophy of Science Part A, 43*(1), 27–37.
Morse, D. J. (2011). *Faith in life. John Dewey's early philosophy*. New York: Fordham University Press.
Mumford, S. (2004). *Laws in nature*. New York: Routledge.
Myers, W. T. (2004). Pragmatist metaphysics: A defense. *Transactions of the Charles S. Peirce Society, 40*(1), 39–52.
Myers, W. T. (2017). Dewey, Whitehead, and process metaphysics. In S. Fesmire (Ed.), *The Oxford handbook of Dewey*. Oxford: Oxford University Press.
Myers, W. T., & Pappas, G. F. (2004). Dewey's metaphysics: A response to Richard Gale. *Transactions of the Charles S. Peirce Society, 40*(4), 679–700.
Nersessian, N. J. (Ed.). (1987). *The process of science. Contemporary philosophical approaches to understanding scientific practice*. Dordrecht: Springer.
Pappas, G. (2014). What difference can "experience" make to pragmatism? *European Journal of Pragmatism and American Philosophy, VI*(2), 200–227.
Pappas, G. F. (2016). John Dewey's radical logic: The function of the qualitative in thinking. *Transactions of the Charles S. Peirce Society, 52*(3), 435–468.
Parrini, P. (2002). *L'empirismo logico. Aspetti storici e prospettive teoriche*. Roma: Carocci Editore.
Peirce (1986). *Writings of Charles S. Peirce: A chronological edition, Volume 3*. Bloomington: Indiana University Press.
Pickering, A. (1992). *Science as practice and culture*. Chicago: The University of Chicago Press.
Pihlström, S. (1999). What shall we do with emergence? A survey of a fundamental issue in the metaphysics and epistemology of science. *South African Journal of Philosophy, 18*(2), 192–210.
Pihlström, S. (2007). The realism issue from a Deweyan perspective. *Americana. E-Journal of American Studies in Hungary, III*(2), 1–28.
Pihlström, S. (2008). How (not) to write the history of pragmatist philosophy of science? *Perspectives on Science, 16*(1), 26–69.
Pihlström, S., Stadler, F., & Weidtmann, N. (Eds.). (2017). *Logical empiricism and pragmatism*. Cham: Springer.
Pihlström, S. (2009). *Pragmatist metaphysics: An essay on the ethical grounds of ontology*. London: Continuum.
Poincaré, H. (1905). *Science and hypothesis*. London: The Walter Scott Publishing Co.
Polanyi, M. (2009). *The tacit dimension*. Chicago: The University of Chicago Press.
Pratt, S. L. (1997). "A sailor in a storm": Dewey on the meaning of language. *Transactions of the Charles S. Peirce Society, 33*(4), 839–862.
Price, H. (2011). *Naturalism without mirrors*. Oxford: Oxford University Press.
Price, H., Blackburn, S., & Brandom, R. (2015). *Expressivism, pragmatism and representationalism*. Cambridge, MA: Cambridge University Press.

Psillos, S. (2001). Is structural realism possible? *Philosophy of Science, 68*(S3), S13–S24.
Putnam, H. (1975). What is mathematical truth? In *Mathematics, matter and method. Philosophical papers, Volume I* (pp. 60–78). Cambridge, MA: Cambridge University Press.
Putnam, H. (2017). Pragmatism and verificationism. In D. MacArthur (Ed.), *Pragmatism as a way of life. The lasting legacy of William James and John Dewey* (pp. 21–35). Cambridge, MA: Harvard University Press.
Putnam, R. A. (2010). Dewey's epistemology. In M. Cochran (Ed.), *The Cambridge companion to Dewey* (pp. 34–54). Cambridge, MA: Cambridge University Press.
Quine, W. V. (1969). Ontological relativity. In *Ontological relativity and other essays* (pp. 26–68). New York: Columbia University Press.
Reichenbach, H. (1938). *Experience and prediction. An analysis of the foundations and the structure of knowledge*. Chicago: The University of Chicago Press.
Reichenbach, H. (1939). Dewey's theory of science. In *The philosophy of John Dewey* (pp. 157–192). New York: Tudor Publishing Co.
Reichenbach, H. (1965). *The theory of relativity and a priori knowledge*. Berkeley: University of California Press.
Reisch, G. A. (2005). *How the cold war transformed philosophy of science*. Cambridge, MA: Cambridge University Press.
Reiss, J. (2015). A pragmatist theory of evidence. *Philosophy of Science, 82*(3), 341–362.
Richardson, A. W. (2002). Engineering philosophy of science: American pragmatism and logical empiricism in the 1930s. *Philosophy of Science, 69*(S3), S36–S47.
Rorty, R. (1998). Dewey between Hegel and Darwin. In *Truth and progress* (pp. 290–306). Cambridge, MA: Cambridge University Press.
Rosenthal, S. B. (1987). The pragmatic a priori: Lewis and Dewey. *The Southern Journal of Philosophy, 25*(1), 109–121.
Rouse, J. (1996). *Engaging science. How to understand its practices philosophically*. Ithaca: Cornell University Press.
Rouse, J. (2002). *How scientific practices matter: Reclaiming philosophical naturalism*. Chicago: Chicago University Press.
Rouse, J. (2015). *Articulating the world. Conceptual understanding and the scientific image*. Chicago: Chicago University Press.
Rubeis, G. (2017). Beyond realism and antirealism? The strange case of Dewey's instrumentalism. In *Logical empiricism and pragmatism* (Vienna circle institute yearbook, pp. 67–82). Cham: Springer.
Ruphy, S. (2017). *Scientific pluralism reconsidered*. Pittsburgh: University of Pittsburgh Press.
Russell, B. (1939). Dewey's new logic. In *The philosophy of John Dewey*. New York: Tudor Publishing Co.
Russell, B. (1940). *An inquiry into meaning and truth*. London: George Allen and Unwin LTD.
Russell, B. (2001). *The problems of philosophy*. Oxford: Oxford University Press.
Ryan, F. X. (1994). Primary experience as settled meaning: Dewey's conception of experience. *Philosophy Today, 38*(1), 29–42.
Ryan, F. X. (1997a). The "extreme heresy" of John Dewey and Arthur F. Bentley I: A star crossed collaboration? *Transactions of the Charles S. Peirce Society, 33*(3), 774–794.
Ryan, F. X. (1997b). The "extreme heresy" of John Dewey and Arthur F. Bentley II: Knowing knowing and the known". *Transactions of the Charles S. Peirce Society, 33*(4), 1003–1023.
Ryan, F. X. (2011). *Seeing together. Mind, matter, and the experimental outlook of John Dewey and Arthur F. Bentley*. Great Barrington, MA: American Institute for Economic Research.
Ryckman, T. (2005). *The reign of relativity: Philosophy in physics 1915–1925*. Oxford: Oxford University Press.
Ryle, G. (1945). Knowing how and knowing that. *Proceedings of the Aristotelian Society, 46*, 1–16.
Saatkamp, H. (Ed.). (1995). *Rorty and pragmatism: The philosopher responds to his critics*. (Vanderbilt library of American philosophy). Nashville: Vanderbilt University Press.

Santarelli, M. (2016). Il dispositivo logico del circuito organico nel pensiero di John Dewey. *Politica.eu, 2*(1), 27–42.
Savage, L. J. (1972). *The foundations of statistics*. New York: Dover Publications Inc.
Schatzki, T. R., Cetina, K. K., & von Savigny, E. (2000). *Practice turn in contemporary theory*. London: Routledge.
Schulkin, J. (2015). *Pragmatism and the search for coherence in neuroscience*. London: Palgrave Macmillan.
Seigfried, C. H. (2001). Pragmatist metaphysics? Why terminology matters. *Transactions of the Charles S. Peirce Society, 37*(1), 13–21.
Seigfried, H. (1993). Dewey's critique of Kant's copernican revolution revisited. *Kant-Studien, 84*(3), 356–368.
Shoemaker, S. (1980). Causality and properties. In P. V. Inwagen (Ed.), *Time and cause* (pp. 109–135). Dordrecht: Springer.
Shook, J. (2002). Dewey and Quine on the logic of what there is. In T. Burke & R. Talisse (Eds.), *Dewey's logical theory. New studies and interpretations* (pp. 93–118). Nashville: Vanderbilt University Press.
Shook, J. R. (1995). John Dewey's struggle with American realism, 1904–1910. *Transactions of the Charles S. Peirce Society, 31*(3), 542–566.
Shook, J. R. (2000). *Dewey's empirical theory of knowledge and reality*. Nashville: Vanderbilt University Press.
Shook, J. R., & Good, J. A. (Eds.). (2010). *John Dewey's philosophy of spirit: With the 1897 lecture on hegel*. New York: Fordham University Press.
Shook, J. R., & Solymosi, T. (Eds.). (2014). *Pragmatist neurophilosophy: American philosophy and the brain*. London: Bloomsbury Publishing.
Slater, M. (2014). *Pragmatism and the philosophy of religion*. Cambridge, MA: Cambridge University Press.
Slater, M. R. (2008). Pragmatism, realism, and religion. *Journal of Religious Ethics, 36*(4), 653–681.
Sleeper, R. W. (2001). *The necessity of pragmatism*. Urbana: University of Illinois Press.
Soler, L., Zwart, S., Lynch, M., & Israel-Jost, V. (Eds.). (2014). *Science after the practice turn in the philosophy, history, and social studies of science*. London: Routledge.
Solymosi, T., & Shook, J. R. (2014). *Neuroscience, neurophilosophy and pragmatism. Brains at work with the world*. London: Palgrave Macmillan.
Stuhr, J. (1992). Dewey's reconstruction of metaphysics. *Transactions of the Charles S. Peirce Society, 28*(2), 161–176.
Stump, D. J. (2015). *Conceptual change and the philosophy of science: Alternative interpretations of the a priori*. New York/London: Routledge.
Sukale, M. (1976). Heidegger and Dewey. In *Comparative studies in phenomenology* (pp. 121–151). The Hague: Springer.
Taylor, H. (2017). Powerful qualities and pure powers. *Philosophical Studies, 175*(6), 1423–1440.
Thayer, H. S. (1952). *The logic of pragmatism. An examination of John Dewey's logic*. New York: The Humanities Press, Inc.
Thayer, H. S. (1990). Dewey and the theory of knowledge. *Transactions of the Charles S. Peirce Society, 26*(4), 443–458.
Tiles, J. E. (1990). *Dewey*. London: Routledge.
van Fraassen, B. C. (2002). *The empirical stance*. New Haven: Yale University Press.
Vetter, B. (2013). Multi-track dispositions. *The Philosophical Quarterly, 63*(251), 330–352.
Votsis, I. (2004). The upward path to structural realism. *Philosophy of Science, 72*(5), 1361–1372.
Welchman, J. (2002). Logic and judgments of practice. In T. Burke, D. M. Hester, & R. B. Talisse (Eds.), *Dewey's logical theory. New studies and interpretations* (chapter 2, pp. 27–42). Nashville: Vanderbilt University Press.
Williams, N. E. (2010). Putting powers back on multi-track. *Philosophia, 39*(3), 581–595.
Wimpenny, J. W. T., & Kinniment, S. (1995). Biochemical reactions and the establishment of gradients within biofilms. In *Microbial biofilms*. Cambridge, MA: Cambridge University Press.

Winther, R. G. (2014). James and Dewey on abstraction. *The Pluralist, 9*(2), 1–28.
Wittgenstein, L. (2009). *Philosophical investigations*. Oxford: Wiley Blackwell.
Worrall, J. (1989). Structural realism: The best of both worlds? *Dialectica, 43*(1–2), 99–124.
Zahar, E. (2001). *Poincare's philosophy: From conventionalism to phenomenology*. Chicago/Lasalle: Open Court Publishing Co.

Index

R. Gronda, *Dewey's Philosophy of Science*, Synthese Library 421,
https://doi.org/10.1007/978-3-030-37562-1

The manufacturer's authorised representative in the EU is Springer Nature Customer Service Centre GmbH, Europaplatz 3, 69115 Heidelberg, Germany. If you have any concerns regarding our products, please contact ProductSafety@springernature.com

Printed and bound by CPI Group (UK) Ltd, Croydon, CR0 4YY
28/09/2026
02224088-0010